my **revision** notes

OCR AS/A-level

GEOGRAPHY

Helen Harris

HODDER EDUCATION
AN HACHETTE UK COMPANY

Although every effort has been made to ensure that website addresses are correct at time of going to press, Hodder Education cannot be held responsible for the content of any website mentioned in this book. It is sometimes possible to find a relocated web page by typing in the address of the home page for a website in the URL window of your browser.

Hachette UK's policy is to use papers that are natural, renewable and recyclable products and made from wood grown in sustainable forests. The logging and manufacturing processes are expected to conform to the environmental regulations of the country of origin.

Orders: please contact Bookpoint Ltd, 130 Park Drive, Milton Park, Abingdon, Oxon OX14 4SE. Telephone: (44) 01235 827720. Fax: (44) 01235 400401. Email education@bookpoint.co.uk Lines are open from 9 a.m. to 5 p.m., Monday to Saturday, with a 24-hour message answering service. You can also order through our website: www.hoddereducation.co.uk

ISBN: 978 1 5104 1551 5

First published in 2017 by
Hodder Education,
An Hachette UK Company
Carmelite House
50 Victoria Embankment
London EC4Y 0DZ

www.hoddereducation.co.uk

Impression number 10 9 8 7 6 5 4 3 2 1
Year 2021 2020 2019 2018 2017

Cover photo ©tr3gi/Fotolia
Typeset in Bembo Std Regular 11/13 by Aptara, Inc.
Printed in Spain

A catalogue record for this title is available from the British Library.

Get the most from this book

Everyone has to decide his or her own revision strategy, but it is essential to review your work, learn it and test your understanding. These Revision Notes will help you to do that in a planned way, topic by topic. Use this book as the cornerstone of your revision and don't hesitate to write in it — personalise your notes and check your progress by ticking off each section as you revise.

Tick to track your progress

Use the revision planner on pages 4 and 5 to plan your revision, topic by topic. Tick each box when you have:

- revised and understood a topic
- tested yourself
- practised the exam questions and gone online to check your answers and complete the quick quizzes

You can also keep track of your revision by ticking off each topic heading in the book. You may find it helpful to add your own notes as you work through each topic.

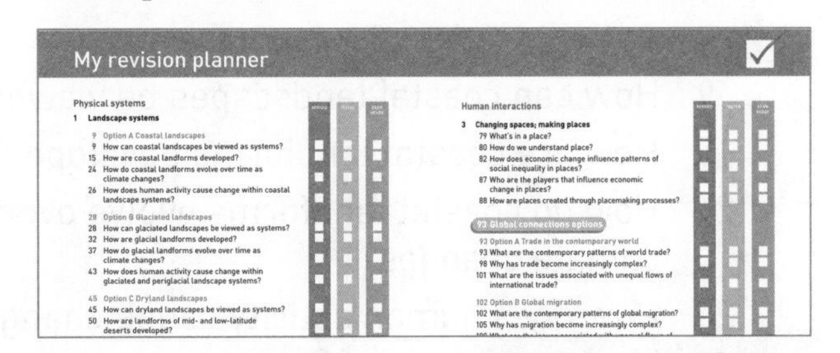

My revision planner

Physical systems

1 Landscape systems

9 Option A Coastal landscapes
9 How can coastal landscapes be viewed as systems?
15 How are coastal landforms developed?
24 How do coastal landforms evolve over time as climate changes?
26 How does human activity cause change within coastal landscape systems?

28 Option B Glaciated landscapes
28 How can glaciated landscapes be viewed as systems?
32 How are glacial landforms developed?
37 How do glacial landforms evolve over time as climate changes?
43 How does human activity cause change within glaciated and periglacial landscape systems?

45 Option C Dryland landscapes
45 How can dryland landscapes be viewed as systems?
50 How are landforms of mid- and low-latitude deserts developed?

Human interactions

3 Changing spaces; making places
79 What's in a place?
80 How do we understand place?
82 How does economic change influence patterns of social inequality in places?
87 Who are the players that influence economic change in places?
88 How are places created through placemaking processes?

93 Global connections options

93 Option A Trade in the contemporary world
93 What are the contemporary patterns of world trade?
98 Why has trade become increasingly complex?
101 What are the issues associated with unequal flows of international trade?

102 Option B Global migration
102 What are the contemporary patterns of global migration?
105 Why has migration become increasingly complex?

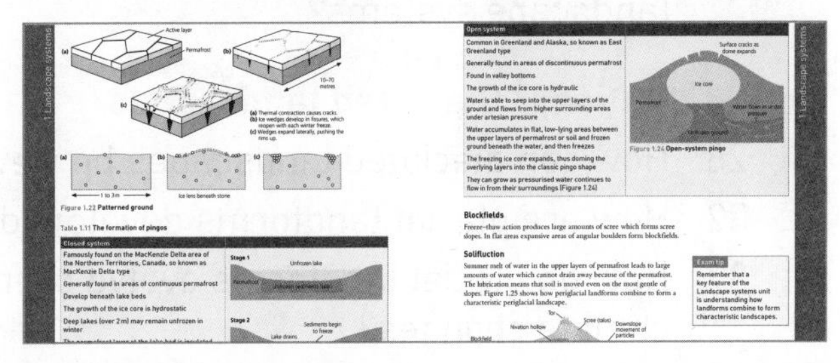

Features to help you succeed

Exam tips and summaries

Expert tips are given throughout the book to help you polish your exam technique in order to maximise your chances in the exam. The summaries provide a quick-check bullet list for each topic.

Typical mistakes

The author identifies the typical mistakes students make and explains how you can avoid them.

Now test yourself

These short, knowledge-based questions provide the first step in testing your learning. Answers are at the back of the book.

Definitions and key words

Clear, concise definitions of essential key terms are provided where they first appear.

Key words from the specification are highlighted in bold throughout the book.

Exam practice

Practice exam questions are provided for each topic. Use them to consolidate your revision and practise your exam skills.

Online

Go online to check your answers to the exam questions and try out the extra quick quizzes at **www.hoddereducation.co.uk/ myrevisionnotesdownloads**

There are also **case studies** at the same address, which are signposted at the end of each chapter.

My revision planner

Physical systems

Human interactions

REVISED | TESTED | EXAM READY

REVISED | TESTED | EXAM READY

4 Geographical debates options

Exam practice answers, quick quizzes and case studies at www.hoddereducation.co.uk/myrevisionnotesdownloads

Countdown to my exams

6–8 weeks to go

- Start by looking at the specification — make sure you know exactly what material you need to revise and the style of the examination. Use the revision planner on pages 4 to 7 to familiarise yourself with the topics.
- Organise your notes, making sure you have covered everything on the specification. The revision planner will help you to group your notes into topics.
- Work out a realistic revision plan that will allow you time for relaxation. Set aside days and times for all the subjects that you need to study, and stick to your timetable.
- Set yourself sensible targets. Break your revision down into focused sessions of around 40 minutes, divided by breaks. These Revision Notes organise the basic facts into short, memorable sections to make revising easier.

REVISED

2–6 weeks to go

- Read through the relevant sections of this book and refer to the exam tips, summaries, typical mistakes and key terms. Tick off the topics as you feel confident about them. Highlight those topics you find difficult and look at them again in detail.
- Test your understanding of each topic by working through the 'Now test yourself' questions in the book. Look up the answers at the back of the book.
- Make a note of any problem areas as you revise, and ask your teacher to go over these in class.
- Look at past papers. They are one of the best ways to revise and practise your exam skills. Write or prepare planned answers to the exam practice questions provided in this book. Check your answers online and try out the extra quick quizzes at **www.hoddereducation.co.uk/myrevisionnotesdownloads**
- Use the revision activities to try out different revision methods. For example, you can make notes using mind maps, spider diagrams or flash cards.
- Track your progress using the revision planner and give yourself a reward when you have achieved your target.

REVISED

One week to go

- Try to fit in at least one more timed practice of an entire past paper and seek feedback from your teacher, comparing your work closely with the mark scheme.
- Check the revision planner to make sure you haven't missed out any topics. Brush up on any areas of difficulty by talking them over with a friend or getting help from your teacher.
- Attend any revision classes put on by your teacher. Remember, he or she is an expert at preparing people for examinations.

REVISED

The day before the examination

- Flick through these Revision Notes for useful reminders, for example the exam tips, summaries, typical mistakes and key terms.
- Check the time and place of your examination.
- Make sure you have everything you need — extra pens and pencils, tissues, a watch, bottled water.
- Allow some time to relax and have an early night to ensure you are fresh and alert for the examinations.

REVISED

1 Landscape systems

Option A Coastal landscapes

How can coastal landscapes be viewed as systems?

Coastal landscapes can be viewed as systems

REVISED

The components of coastal landscape systems including energy flows

The system approach is a way of analysing the relationships within a unit, e.g. a coast. It consists of several components (stores) and processes (links) that are connected and represented in a flow diagram.

The coast is an **open system** (Figure 1.1) meaning that energy and matter can cross the boundary of the system to the surrounding environment. It has:

- **inputs:** which include **kinetic energy** from waves and wind, **thermal energy** from the heat of the sun, **potential energy** from material on slopes and material from **processes** of weathering, mass movement, erosion and deposition
- **outputs:** which include marine and wind erosion from beaches
- **throughputs:** stores including beach sediment and flows (transfers) such as the movement along a beach by longshore drift

The combination of these factors forms distinctive landscapes that are made up of a range of erosional and depositional landforms created by natural geomorphic processes and reflecting human activity.

When the inputs and outputs of a system are equal it is in a state of **equilibrium**. However, coasts are dynamic (constantly changing) places and the equilibrium is often disturbed, resulting in **dynamic equilibrium**.

Change occurs to upset the balance of the system; for a coast this may be landslides, storms or human activity, for example. The system adjusts by a process of **feedback**, which can be either **positive** (an initial change bringing about further change in the same direction) or **negative** (the system is returned to its normal functioning).

Revision activity

Produce a summary diagram including the terms in bold in this section and a definition for each.

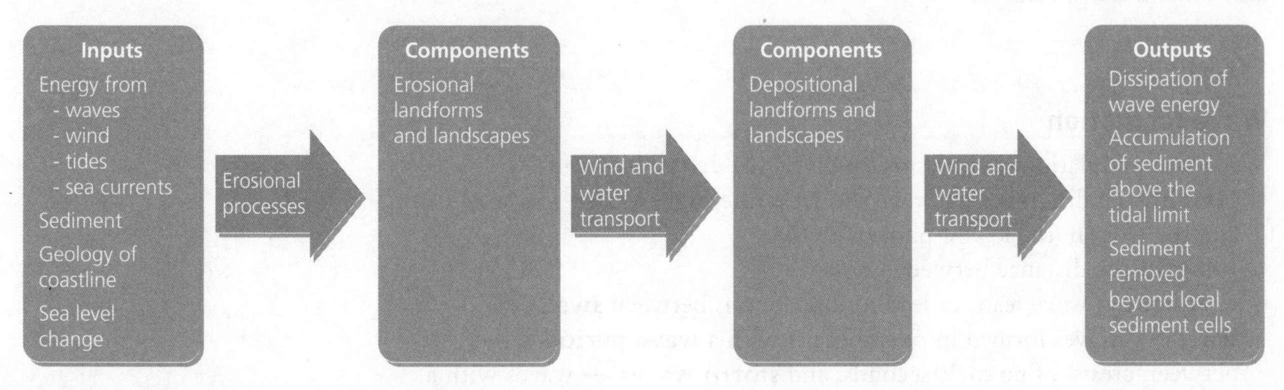

Figure 1.1 The coast as an open system

Sediment cells

Sediment movement occurs in distinct areas called **cells** — a stretch of coastline within which the movement of sediment, sand and shingle is largely self-contained (Figure 1.2). If part of a larger cell they are called **sub-cells**. For example, the **Flamborough Head–Humber Estuary** sub-cell is part of the larger **Flamborough Head–The Wash** cell.

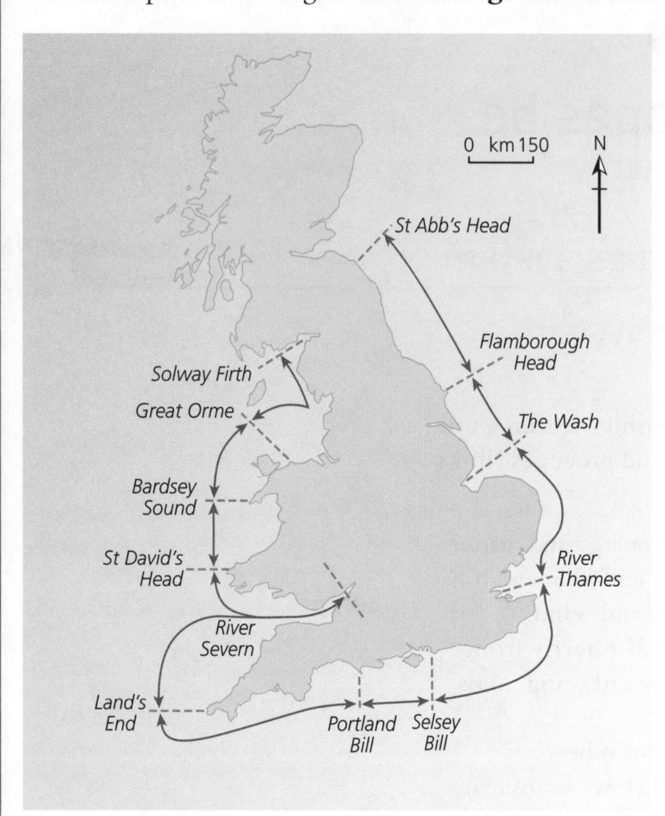

Figure 1.2 **Coastal sediment cells around England and Wales**

Coastal landscape systems are influenced by a range of physical factors

REVISED

A range of physical factors affect the way coastal processes work and thereby the shaping of the landscape. These factors are often interrelated and vary over space and time.

Wind

Wind is the primary source of energy for a range of other processes, e.g. erosion and transportation (Figure 1.3).

Waves

Wave formation

Waves are undulations on the surface of the sea driven by wind.

- **Height:** the difference between the crest (the highest part) of a wave and the trough (the lowest part) of a wave.
- **Length:** the distance between crests.
- **Frequency:** wave features lead to distinctions between **swell waves** — waves formed in open oceans with a **wave period** (time between crests) of up to 20 seconds, and **storm waves** — waves with a short length, greater height and wave period of up to 5 seconds.

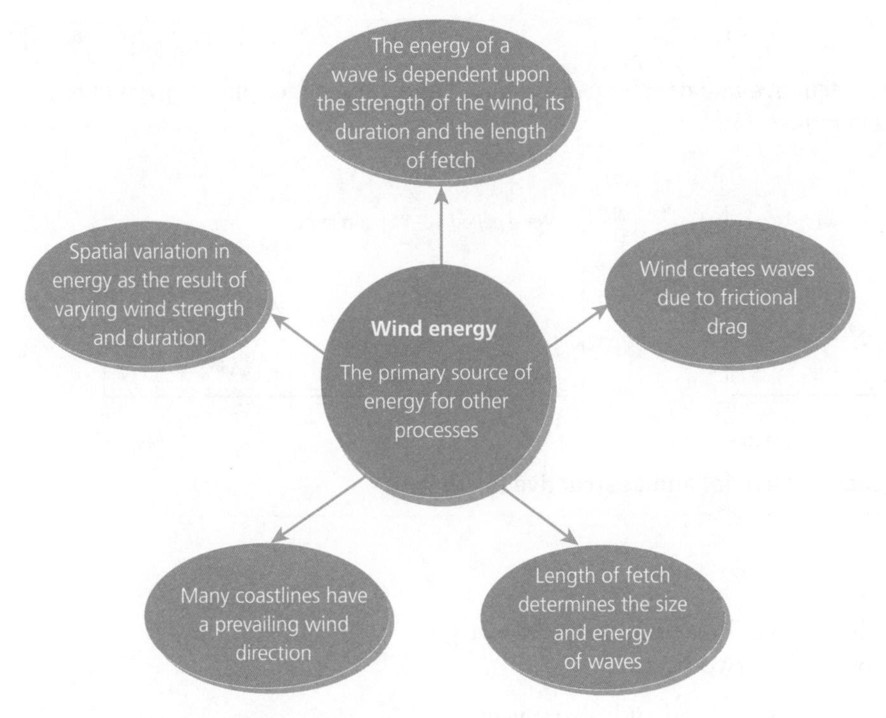

Figure 1.3 Wind energy in coastal systems

Wave development and breaking

A wave enters shallow water → friction with the seabed increases → the wave slows as it drags along the bottom → the wavelength decreases and successive waves start to bunch up → the wave increases in height → and plunges or breaks onto the shoreline.

Breaking waves can be **spilling** (steep waves on gently sloping beaches), **plunging** (steep waves on steep beaches) or **surging** (low-angle waves on steep beaches).

The wash of water up the beach is the **swash**; the drag back down the beach is the **backwash**.

Constructive and destructive waves

- **Constructive:** low, long length (up to 100 m), low frequency (6–8 per minute), gentle spill onto the shore. The swash loses volume and momentum, leading to a weak backwash, and sediment movement off the beach is low. Swash energy exceeds backwash energy. Material is slowly and gradually moved up the beach. Forming: **berms**.
- **Destructive:** high, steep, high frequency (10–14 per minute), rapid approach to shoreline, little forward movement of the water, powerful backwash; sediment is pulled away from the beach. Swash energy is less than backwash energy. Very little material is moved up the beach. Forming: **storm beaches**.

Typical mistake

Do not assume that destructive waves create erosional features and constructive waves depositional features. The description of the waves refers to the transportation of sediment in the coastal zone, i.e. constructive waves deposit sediment and destructive waves remove it.

Revision activity

Draw a simplified diagram of a constructive and destructive wave and apply the descriptions given in this section. Use the outline provided in Figure 1.4.

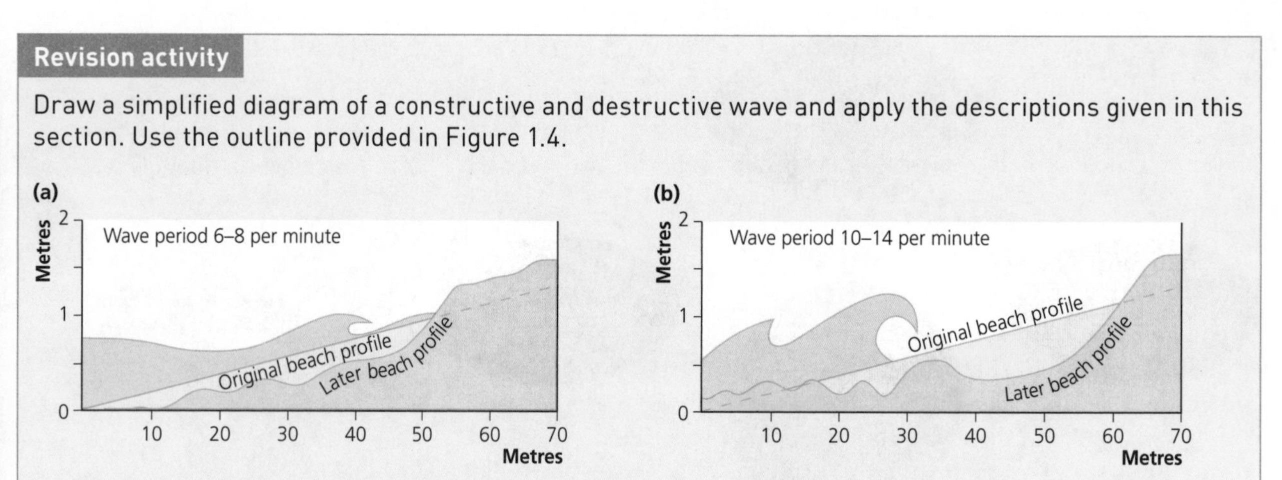

Figure 1.4 Outline diagram of a constructive (a) and destructive (b) wave

Wave refraction

Wave refraction is the process by which waves break onto an irregularly shaped coastline, e.g. a headland separated by two bays.

Waves drag in the shallow water approaching a headland → the wave becomes high, steep and short → the part of the wave in the deeper water moves forward faster → the wave bends → the low-energy wave spills into the bays as most of the wave energy is concentrated on the headland.

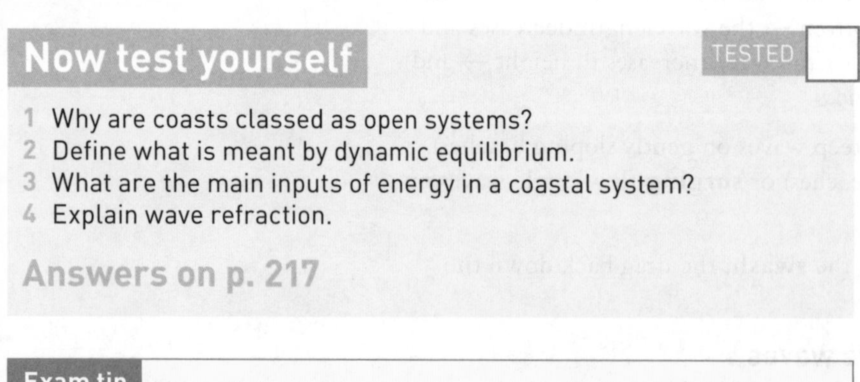

Now test yourself TESTED

1 Why are coasts classed as open systems?
2 Define what is meant by dynamic equilibrium.
3 What are the main inputs of energy in a coastal system?
4 Explain wave refraction.

Answers on p. 217

Exam tip

In physical geography learning the correct sequence to an explanation is key to achieving accuracy.

Tides

Tidal cycles

The periodic rise and fall of the sea surface is produced by the gravitational pull of the moon and (to a lesser extent) the sun.

The moon pulls water towards it, creating a high tide → there is a compensating 'bulge' on the opposite side of the Earth → at locations between the two bulges there is a low tide → as the moon orbits the Earth the high tides follow it → the highest tides occur when the moon, Earth and sun are all aligned and so the gravitational pull is strongest = **spring tides** with a high tidal range → when the moon and sun are at angles to each other the gravitational pull is weak = **neap tides** with a low tidal range.

Tidal range

Tidal range is a significant factor in the development of coastlines as it influences where wave action occurs, the weathering processes and the impact of processes between tides, such as scouring.

Tidal range is the vertical difference in height between the consecutive high and low water over a tidal cycle.

Geology

Lithology

Lithology refers to the chemical and physical structure of rocks. This will impact on physical processes such as weathering, mass movement and erosion. Weak rocks such as clay will erode faster than resistant rocks such as basalt. Chalk and limestone are susceptible to chemical weathering because of their calcium carbonate content which is soluble in weak acids.

Structure

Structure refers to features of jointing, faulting and bedding planes in rocks and also to their **permeability**. Permeable rocks include chalk (water absorbed through tiny pores) and limestone (water absorbed through joints).

Permeability is the ability of rock to absorb water through joints and pores.

Structure also affects the shape of the coastline: where rocks lie parallel to the coastline it tends to be straight or **concordant**; where rocks lie at right angles to the coast a series of headlands and bays are formed according to the location of weak or resistant rock — this is a **discordant** coastline. See Figure 1.5.

Structure also affects the 'dip' of rocks towards the coastline: landward-dipping rock layers lead to steep cliffs and for seaward-dipping rock layers cliffs follow the angle of the dip.

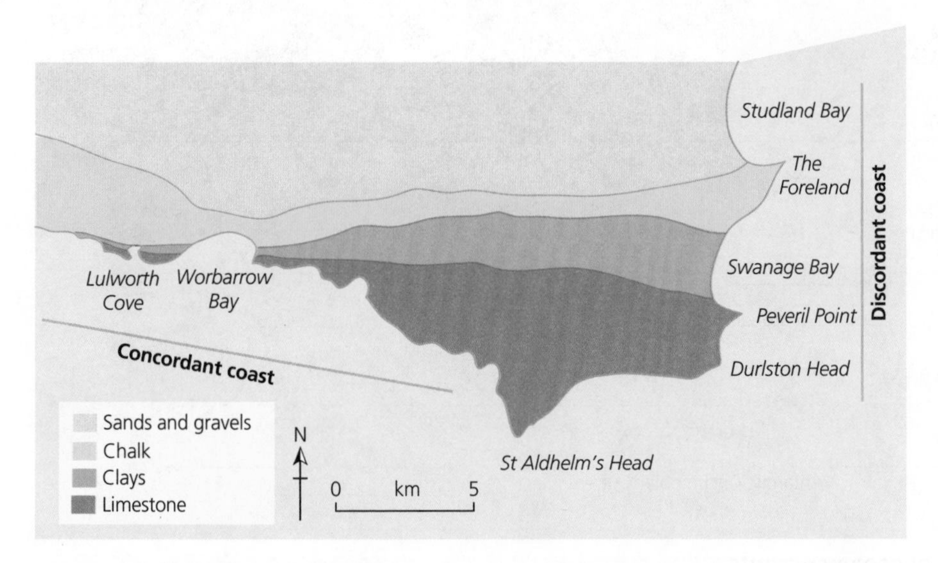

Figure 1.5 The Isle of Purbeck: east-facing discordant and south-facing concordant coastline

Ocean currents

Currents are the permanent or seasonal movement of water in the seas and oceans. There are three types (Table 1.1).

Table 1.1 Types of currents

Longshore currents
Most waves approach the shoreline at an angle. This creates a current of water running parallel to the shoreline Effect: transports sediment parallel to the shoreline
Rip currents
These are strong currents moving away from the shoreline as a result of a build-up of seawater and energy along the coastline Effect: hazardous for swimmers
Upwelling
The global pattern of currents circulating in the oceans can cause deep, cold water to move towards the surface, displacing the warmer surface water Effect: a cold current rich in nutrients

Global pattern of ocean currents

The global pattern of ocean currents is generated by the Earth's rotation and the currents are set in motion by the wind. Warm ocean currents transfer heat from low latitudes to high latitudes and cold ocean currents from high to low latitudes. This transfer of heat energy is significant to coastal development as it affects air temperature and, therefore, **sub-aerial processes**. Figure 1.6 shows the global pattern of ocean currents.

Sub-aerial processes is a collective term for weathering and mass movement processes.

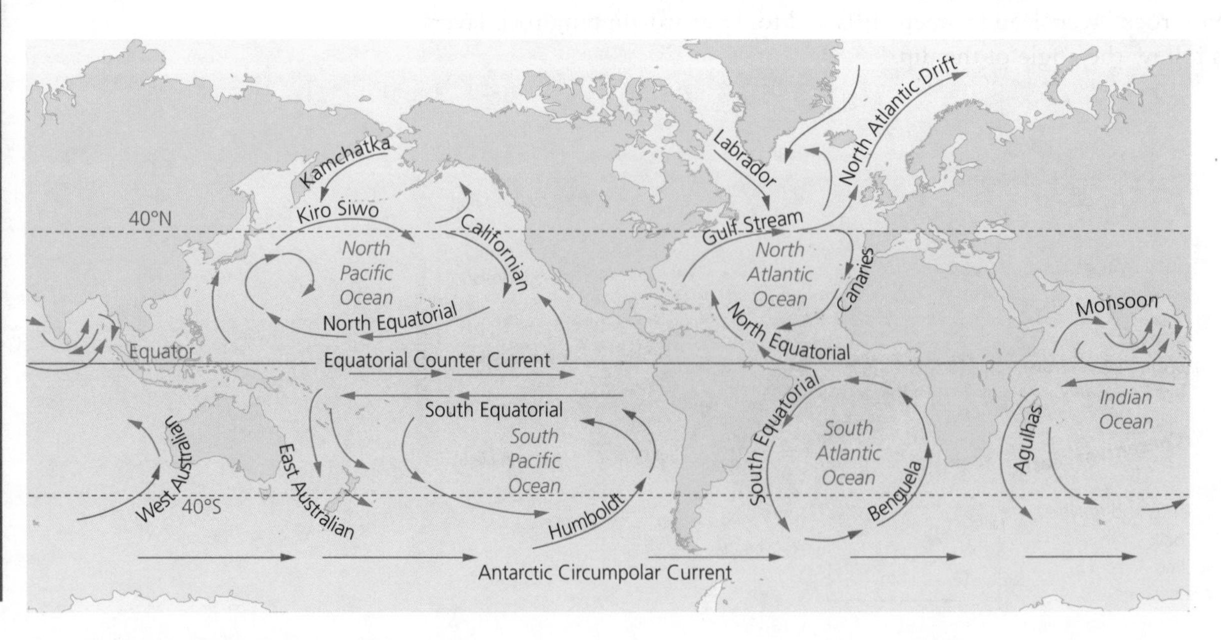

Figure 1.6 The global pattern of ocean currents

Exam tip

Be very clear on the dominant processes for different locations, e.g. tidal energy dominates in estuarine environments, wind energy in dune environments. Consider the geographical context of a location/example.

Coastal sediment is supplied from a variety of sources

REVISED

Coastal sediments form depositional features such as beaches and mudflats, and on coastlines there is a delicate balance between the input and removal of sediment, which is referred to as a sediment budget. In its simplest form:

- more material added than removed = a positive budget (accretion of material) → shoreline builds to the sea
- more material removed than added = a negative budget → shoreline recedes landward

Calculating sediment budgets is complex, as all possible inputs, stores (sinks) and outputs of sediment need to be identified.

Revision activity

The notes in this section on wind, waves, tides, geology and currents illustrate different methods of making revision notes:

- a spider diagram
- notes with bullet points and bold or colour print for key terms
- a table summary
- annotated diagrams

Choose one of these methods to make your own revision summary of the sources of coastal sediment.

Include the following:

- terrestrial sources (fluvial, weathering, mass movement, aeolian deposits, longshore drift)
- offshore sources (e.g. marine deposits)
- human sources (e.g. beach nourishment)

Typical mistake

River and seabed sources account for the highest proportion of sediment sources, not cliff erosion as is often thought.

How are coastal landforms developed?

Coastal landforms develop due to a variety of interconnected climatic and geomorphic processes

REVISED

Geomorphic processes

Figure 1.7 provides a summary of coastal processes.

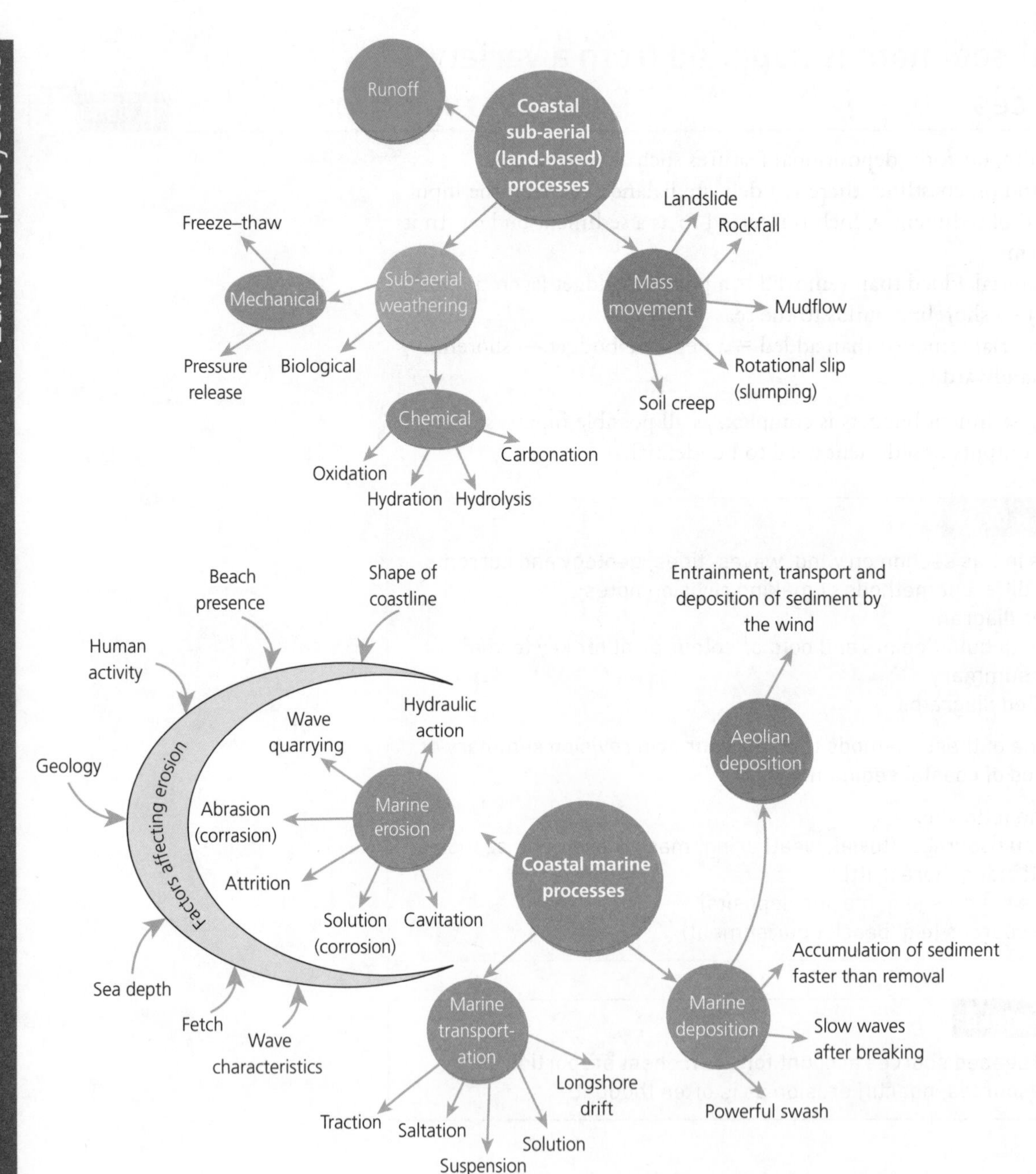

Figure 1.7 Coastal processes

Typical mistake

When referring to coastal processes in the formation of landforms and landscapes, don't just say 'erosion processes' — be specific and state what type, e.g. hydraulic action or abrasion.

Revision activity

Complete a table describing the different weathering processes in coastal landscapes. Table 1.2 shows an example of structure. Colour-code the different categories and then in pairs test each other by giving a term and asking for an explanation or give an explanation and ask for the term to which it refers.

Weathering processes

Weathering is a significant process in the formation of coastal landscapes. There are three types of weathering: physical (or mechanical), chemical (involving chemical reactions) and biological (the result of plant and animal activity).

Table 1.2 Weathering processes in coastal landscapes

Mechanical weathering	
Freeze–thaw	
Pressure release	
Thermal expansion	
Salt crystallisation	
Chemical weathering	
Oxidation	
Carbonation	
Solution	
Hydrolysis	
Hydration	
Biological weathering	
Tree roots	
Organic acids	

Mass movement processes

These processes refer to the movement of material (**regolith**) down a slope. They are sub-aerial (above ground) and are dependent on slope angle, particle size, temperature and saturation. The main forms of mass movement in coastal areas are summarised in Table 1.3.

Regolith is a loose layer of rocky material lying over bedrock.

Table 1.3 Processes of mass movement in coastal environments

Landslides	Cliffs made of softer rocks slip when lubricated by rainfall
Rockfalls	Rocks undercut by the sea or slopes affected by mechanical weathering
Mud flows	Heavy rain causes fine material to move downhill
Rotational slip/slumping	Where soft material overlies resistant material and excessive lubrication takes place
Soil creep	Very slow movement of soil particles down slope

Wave processes

Breaking waves erode the coastline through a range of processes summarised in Table 1.4. Waves can also supply sediment to the system which is either deposited or transported (Table 1.5).

Table 1.4 Processes of wave erosion in coastal environments

Name of process	Description
Hydraulic action	Wave pounding — the force of the water on the rocks
Wave quarrying/pounding	Breaking wave exerts pressure on the rock; air is trapped in cracks in a cliff face; as the water pulls back air is released under pressure. The rock face is weakened over time
Abrasion/corrasion	Sand, shingle and boulders picked up by the sea and hurled against a cliff
Attrition	The wearing down of rocks and pebbles as they rub against each other, making them smaller and rounder
Solution/corrosion	Where fresh water mixes with salt water acidity may increase and carbon-based rocks will be broken down

Exam tip

It does not matter which term you use — abrasion/corrasion or solution/corrosion. Best practice is to pick one, learn it and stick to it.

Longshore drift is the process by which sediment deposited on the shore is moved along the shoreline

Table 1.5 Processes of transportation by waves in coastal environments

Name of process	Description
Traction	Large boulders rolled along the sea bed
Saltation	Small stones bounce along the sea bed
Suspension	Very small particles carried in moving water
Solution	Dissolved material
Longshore drift	Waves approach the shore at an angle; swash moves material up the beach in the same direction as the wave; backwash moves the material back down the steepest gradient — usually perpendicular to where it is picked up by the next incoming wave

Deposition occurs when velocity and/or volume of water decreases and energy is reduced. Deposition takes place in coastal environments when:

- sediment accumulation exceeds removal
- waves slow after breaking
- backwash water percolates into beach material
- there is a sheltered area such as an estuary

Fluvial processes

These are significant in estuarine environments. Fluvial erosion, weathering and mass movement processes supply sediment to river channels. This is then transported downstream and deposited as rivers enter the sea.

Mud flats and **salt marshes** are landforms that form in sheltered low-energy coastlines. They are associated with large tidal ranges where powerful currents transport large quantities of fine sediment.

Aeolian processes

Coastal landscapes are significantly influenced by winds. Wind picks up sand particles and moves them, a process known as **deflation**. Attrition on land by windblown particles is also effective over long distances. When the wind speed falls, material carried by the wind will be deposited.

Aeolian processes are erosion, transport and deposition by the wind.

Now test yourself TESTED

5 Explain the transfer of heat by ocean currents.
6 What is a sediment budget and how is it calculated?
7 Why are weathering processes an important factor in the development of coastal landscapes?

Answers on p. 217

The formation of distinctive erosional landforms

Headlands and bays

Headlands and bays form where there are bands of different rock with different resistance to erosion. Where these bands of rock lie perpendicular to the coastline the weaker rock erodes more quickly forming a bay and the hard rock is left as a headland — this is a **discordant** coastline. Where the bands of rock lie parallel to the coastline the hard rock lies on the seaward side of the coastline and bays develop when a weakness is eroded landward — this is a **concordant** coastline. Figure 1.5 on p. 13 shows sections of discordant and concordant coastlines.

Exam tip

Landforms are small-scale features, while landscapes refer to the way features interconnect to form a landscape.

Cliffs and wave-cut platforms

The formation of cliffs and wave-cut platforms is outlined in Figure 1.8.

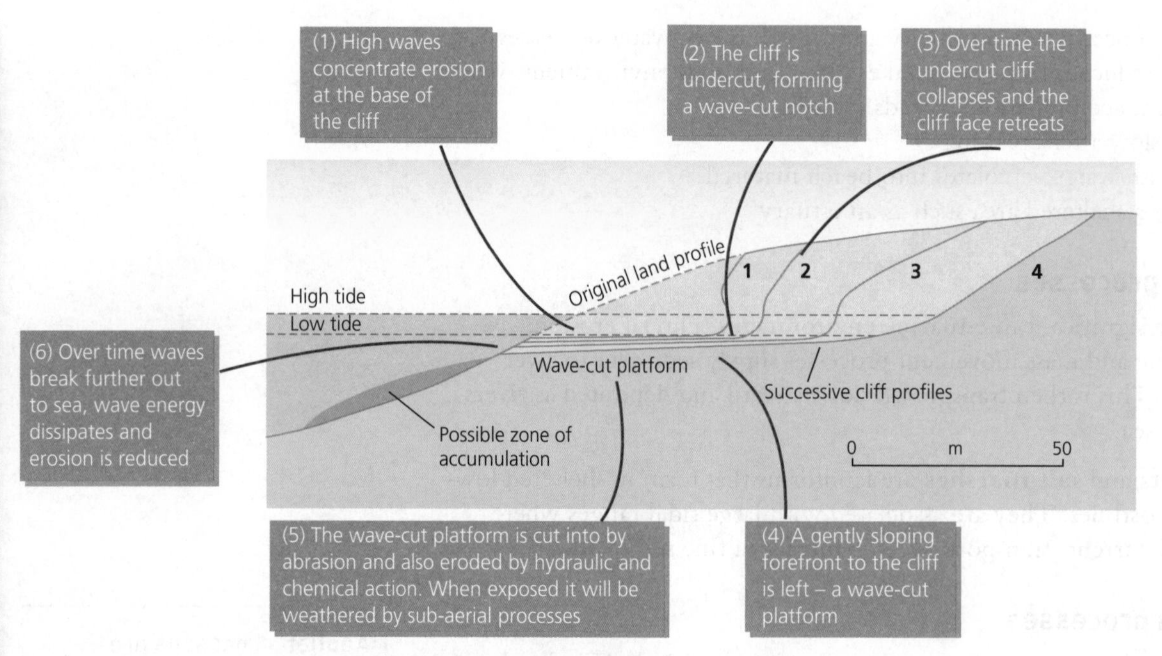

Figure 1.8 The formation of cliffs and wave-cut platforms

Exam tip

Sometimes in exam questions you might be asked to *apply* your knowledge to unfamiliar contexts or locations. Make sure that you can apply reasoning of how different factors will affect different coastlines.

Geos and blowholes

Geos are narrow, steep-sided inlets formed on a coastline where there is a weakness (joint or fault) in the rock which is exposed by erosion processes such as hydraulic action. If a geo becomes enlarged by continual erosion and the roof collapses then it becomes a blowhole, appearing as a vertical shaft that reaches the cliff top.

Caves, arches, stacks and stumps

The formation of caves, arches, stacks and stumps is outlined in Figure 1.9.

Exam tip

Not all cliffs develop caves, arches and stacks. Geology can have an important effect, e.g. clay may form cliffs but will not support caves, arches and stacks as chalk would.

Revision activity

Practise drawing an annotated sketch diagram to explain the formation of cliffs and wave-cut platforms or caves, arches and stacks. Base your diagram on a located example.

Now test yourself

TESTED

8 What factors affect coastal landscapes and their characteristic landforms?
9 Describe the *characteristics* of a landscape of coastal erosion.
10 List the processes involved in the formation of caves, arches and stacks.

Answers on p. 217

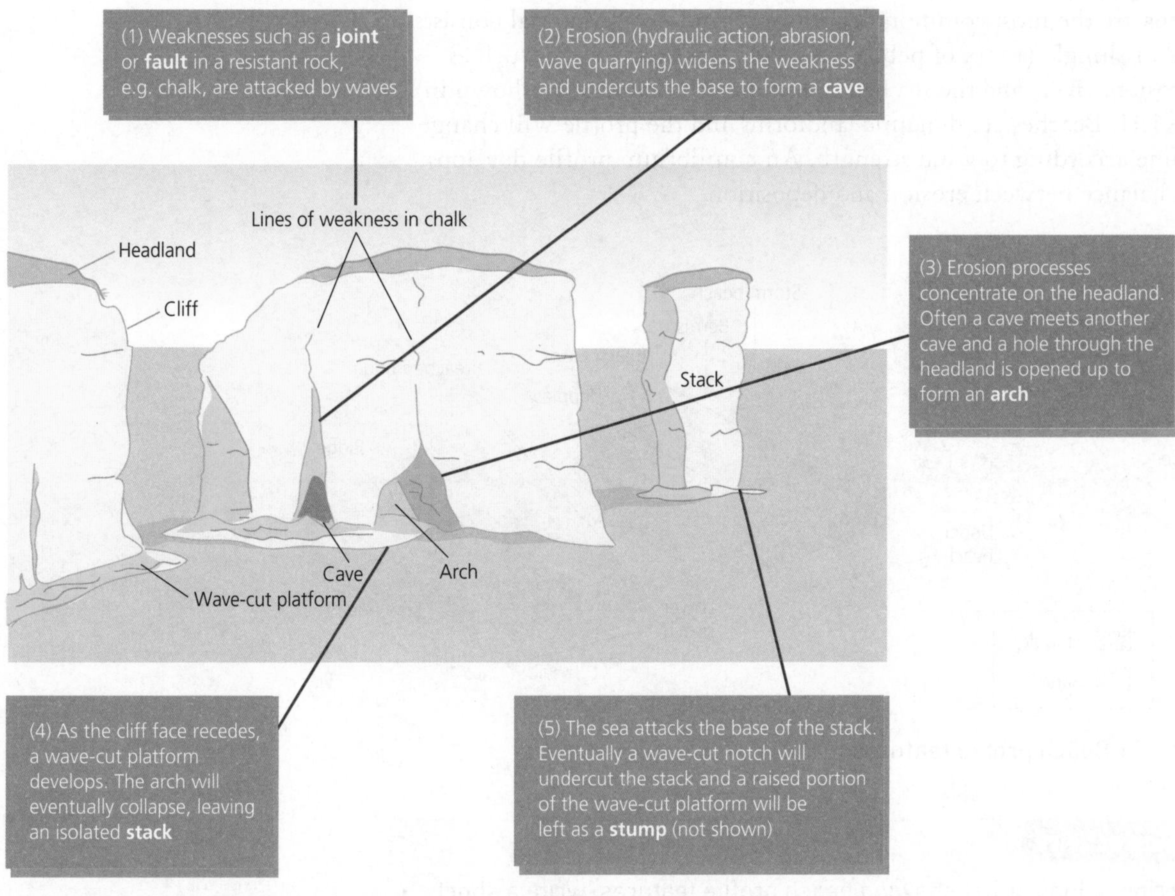

Figure 1.9 The formation of caves, arches, stacks and stumps

The formation of distinctive depositional landforms

Beaches, spits, tombolos, onshore and offshore bars

The landforms formed by coastal deposition are summarised in Figure 1.10 as a coastal landscape.

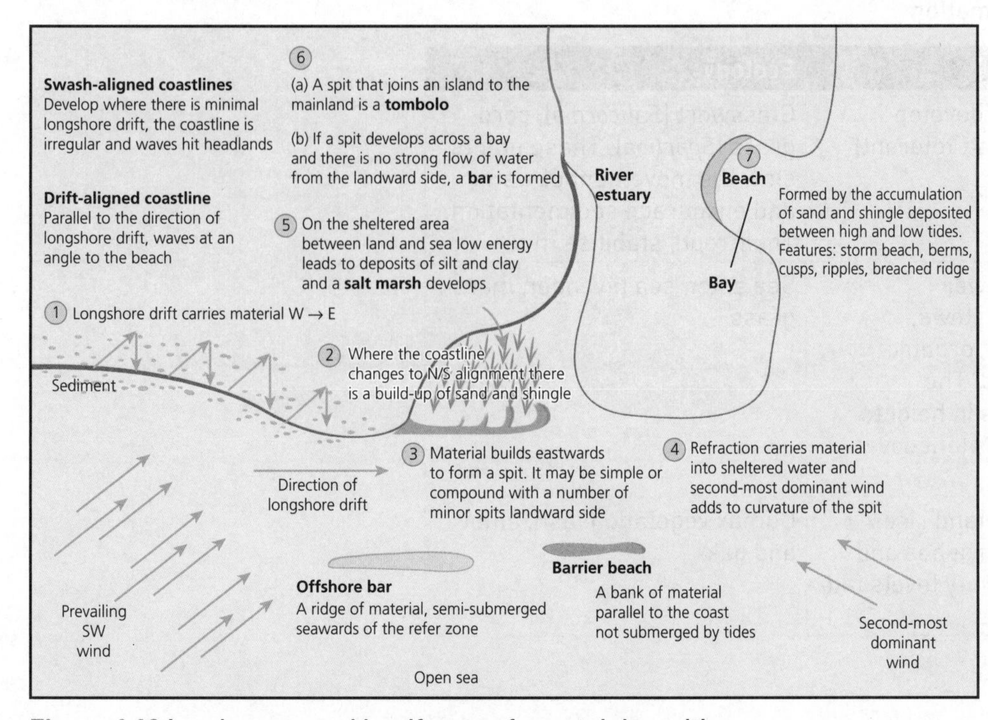

Figure 1.10 Landscapes and landforms of coastal deposition

Beaches are the most common depositional landform. Material consists of sand or shingle (a mix of pebbles and cobbles) from sources such as cliff erosion, rivers and the sea bed. Beach profile features are shown in Figure 1.11. Beaches are dynamic landforms and the profile will change over time according to wind strength. An equilibrium profile develops with a balance between erosion and deposition.

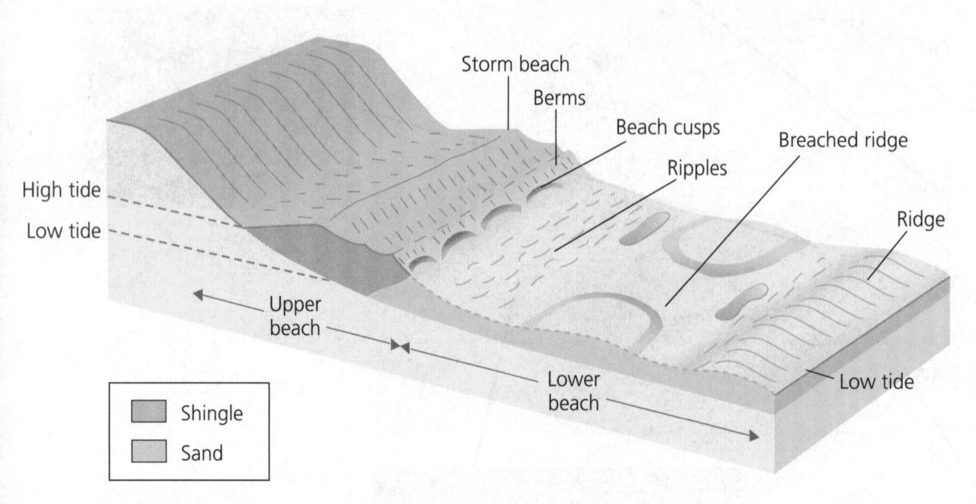

Figure 1.11 Beach profile features

> **Revision activity**
>
> Referring to Figure 1.11 showing beach profile features, write a short sentence to explain the following: storm beach, berms, cusps, ripples, ridge and breached ridge.

Salt marshes form a flat landscape in low-lying estuarine areas. They form over time in three distinct stages, each showing a specific ecology (Table 1.6).

Table 1.6 Salt marsh formation

Stage	Description	Ecology
1	Pioneer species develop (halophytes — salt tolerant)	Glasswort (*Salicornia*), cord grass (*Spartina*). These plants slow the movement of water and encourage sedimentation. Their roots stabilise mud
2	Soil develops, lower salinity, current slows, more deposition, organic matter produced. The marsh increases in height. Biodiversity and plant cover increase	Sea aster, sea lavender, marsh grass
3	Mud level rises, land rises above sea level, rushes and reeds grow. Salinity levels fall and soil develops	Climax vegetation: ash, alder and oak

Now test yourself

TESTED

11 What is the difference between the formation of flat and steep beach profiles?
12 How does vegetation change as a salt marsh develops?
13 What are cusps?

Answers on p. 217

Revision activity

Relate the formation of the deposition features outlined in this section to a located example, e.g. Orford Ness, East Anglia. Base your summary on a sketch map similar to Figure 1.10.

Coastal landforms are inter-related and together make up characteristic landscapes

REVISED

Low-energy and high-energy coasts

Figure 1.12 summarises the features of high- and low-energy coastlines.

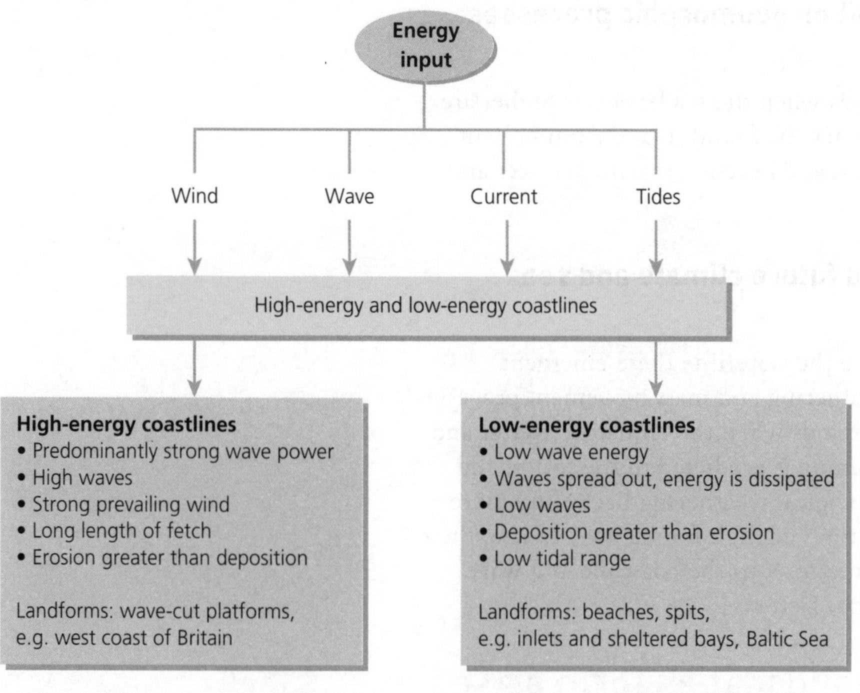

Figure 1.12 The features of high- and low-energy coastlines

Case studies of one high-energy coastline and one low-energy coastline

Online (see p. 3) you will find a case study on the Nile Delta to illustrate a low-energy coastline.

Revision activity

Based on the example of the Nile Delta, make your own revision summary of a high-energy coastline studied in class.

How do coastal landforms evolve over time as climate changes?

Emergent coastal landscapes form as sea level falls

REVISED

Eustatic change is a global change in the volume of sea water resulting from a rise or fall in the level of the sea itself, e.g. caused by the retreat of ice following a glacial period. Eustatic changes are influenced by variations in mean global temperature. **Isostatic change** is a local change in sea level resulting from the land rising or falling relative to the sea, e.g. tectonic movements.

Exam tip

It is important to know the factors that can affect global temperature change and the volume of water in oceans, e.g. variations in the Earth's orbit and tilt.

Cooling climate and emergent landscapes

Climate change and sea level fall

Climate change can lead to sea level fall in the following way.

Fall in global temperature → more precipitation in the form of snow → snow turns to ice → more water stored on the land as solid ice rather than liquid water which is returned to the oceans.

The influence of sea level fall on geomorphic processes and landforms

Landforms shaped by wave processes when the sea level was higher are exposed when sea level falls. They may be found inland from present coastlines. Such landforms include raised beaches, marine terraces and abandoned cliffs. See Figure 1.13.

Modifications by present and future climate and sea level change

As they are now exposed and above the waterline these emergent landforms are now affected by weathering and mass movement processes, e.g. freeze–thaw. In post-glacial periods when the climate is wetter and warmer, vegetation develops. Warming is predicted in the future and this could lead to chemical and biological weathering becoming more influential. If the rise in temperature is enough for sea level to rise, these emergent features may then become closer to the coastline and wave processes will again be an influential factor.

Submergent coastal landscapes form as sea level rises

REVISED

Warming climate and submergent landscapes

Climate change and sea level rise

Climate change can lead to sea level rise in the following way.

Rise in global temperature → melting of ice stores on the land → increase in the volume of water in the oceans → sea level rise. A 1°C rise in mean global temperature could lead to a rise of 2 m at sea level.

The influence of sea level rise on geomorphic processes and landforms

The main influence of a rise of sea level is the submergence of features such as river valleys, forming **rias**, and glacial valleys, forming **fjords**. **Shingle beaches** also appear where former coastal sediment is pushed onshore by wave action. See Figure 1.13.

Modifications by present and future climate and sea level change

Rias and fjords can be modified by wave processes; the height and intensity of the waves will be increased if there are more storm conditions as a result of future climate change. The valley sides will be modified by sub-aerial weathering processes, e.g. mechanical weathering. Shingle beaches will be modified by processes that transport sediment such as **longshore drift**.

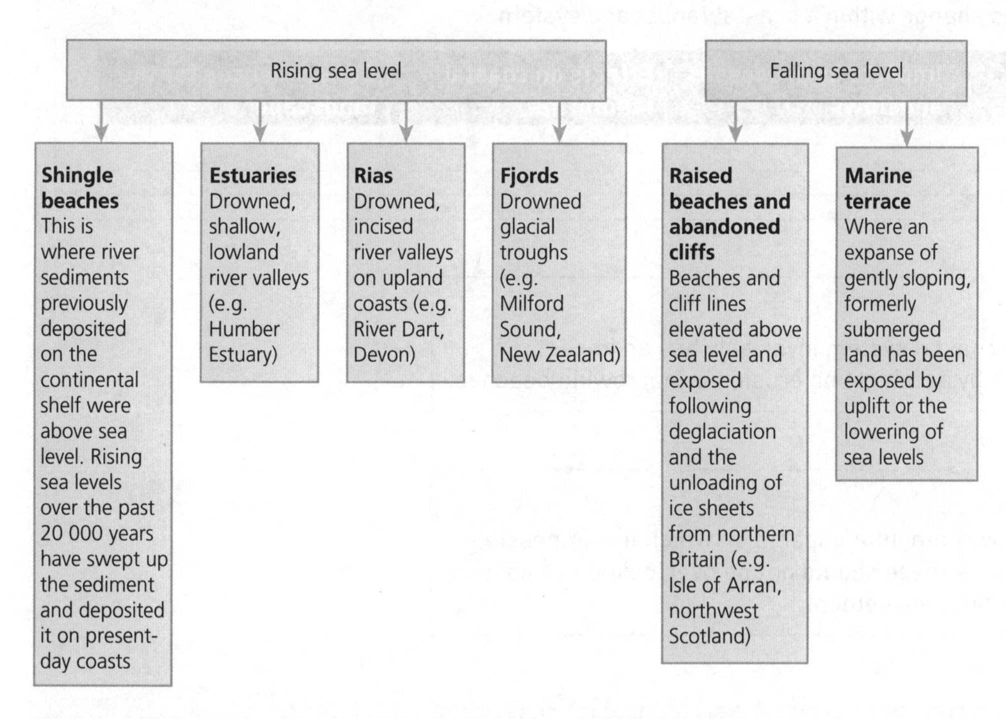

Figure 1.13 Changing sea level and coastal landforms

Now test yourself

TESTED

14 What are the main features of a high-energy coastline?
15 What landforms characterise a low-energy coastline?
16 How are abandoned cliffs formed?

Answers on p. 217

How does human activity cause change within coastal landscape systems?

Human activity intentionally causes change within coastal landscape systems

REVISED

Case study of **one** coastal landscape that is being managed

Revision activity

From the located example you have studied in class, copy and complete this table to summarise how human activity causes change within a coastal landscape system.

Management strategy	Impacts on processes and flows	Effects on coastal landforms	Consequences for the landscape

Exam tip

Any intervention in the coastal system invariably has an impact elsewhere, for example by accelerating erosion or narrowing beaches.

Typical mistake

There are a range of environmental impacts on which it is impossible to place monetary value — these should not be overlooked in a cost-benefit analysis of coastal management.

Economic development unintentionally causes change within coastal landscape systems

REVISED

Case study of **one** coastal landscape that is being used by people

Revision activity

Select a suitable method of presentation, e.g. star diagram or annotated map, and summarise the case study covered in class for a coastal landscape being used by people. Focus on the following:

- the economic development taking place and the reasons for it
- the unintentional impacts on processes and flows
- the effect of the impacts in changing coastal landforms
- the consequences of these changes on the landscape

Exam tip

Case study examples need careful revision of accurate, up-to-date facts, if evaluation is to be addressed in a high-level response.

Summary

- The coastal system is an open, dynamic system driven by flows of energy.
- The sources and flows of sediment are key sub-systems.
- Physical factors such as wind, waves, tides, geology and currents affect coastal landscapes.
- Geomorphological, marine and sub-aerial processes lead to the formation of characteristic coastal landforms which combine and integrate to form coastal landscapes unique to location and geographical contexts.
- Rising and falling sea levels impact on coastal landscapes.
- Human activity can intentionally cause change within coastal landscape systems.
- Economic development can unintentionally cause change within coastal landscape systems.

Exam practice

1 Using evidence from Figure 1.6, p. 14, suggest why warm ocean currents have a greater effect on coastal landscapes than cold ocean currents. [3]

2 Suggest reasons for a predominance of erosional features on the SW coast of the UK. [4]

3 Explain the formation of salt marshes. [8]

4 Explain the role of sub-aerial processes in the formation of coastal landscapes. [8]

5 Using the data in Table 1.7:

(a) Calculate the interquartile range (IQR). [2]

(b) A fieldwork study of the relationship between pebble size and distance from the sea on beach A gave a Spearman rank correlation result of R_s–0.843 for 18 paired values. Using Table 1.8, state the null hypothesis and interpret this result. [4]

Table 1.7 Pebble sample, beach A

Sample number	1	2	3	4	5	6	7	8	9	10	11
A-axis (cm)	2.5	3.1	4.6	2.2	1.9	4.8	5.2	6.2	3.3	4.2	4.7

Table 1.8

Degrees of freedom	0.05 significance level	0.01 significance level
17	±0.412	±0.582
18	0.399	0.564
19	0.388	0.549

Answers and quick quiz 1A online

ONLINE

Option B Glaciated landscapes

Glaciated landscapes have been formed to a greater or lesser extent by the action of glaciers. Cold climates supporting glaciers occur in both locations of high latitude — Antarctica and Greenland. Mountain ranges with glaciers include the Himalayas. They also include areas glaciated in the past, e.g. northern Britain. Landscapes are usually classified as erosional or depositional.

Exam tip

There are a range of locations and types of glaciated landscapes. Be aware of this when responding to exam questions.

How can glaciated landscapes be viewed as systems?

Glaciated landscapes can be viewed as systems

REVISED

The system approach is a way of analysing the relationships within a unit. It consists of several components (stores) and processes (links) that are connected and can be represented in a flow diagram.

Glaciated landscape systems store and transfer energy and material on timescales that can vary from a few days to millennia.

The components of glaciated landscape systems including flows of energy

Glaciated landscape systems are **open systems** (Figure 1.14), meaning that energy and matter can cross the boundary of the system to the surrounding environment. They have:

- **inputs:** kinetic energy from wind and the movement of glaciers, thermal energy from the heat of the sun, and potential energy from material on slopes and material from **processes** of weathering, mass movement and deposition
- **outputs:** which include glacial and wind erosion from rock surfaces; evaporation, sublimation and meltwater
- **throughputs:** stores including ice, water and debris accumulations and flows (transfers) such as the movement of ice, water and debris downslope under gravity

The combination of these factors forms distinctive landscapes which are made up of a range of erosional and depositional landforms created by natural geomorphic processes and reflecting human activity.

When the inputs and outputs of a system are equal it is in a state of **equilibrium**. If the rate at which snow and ice are added to a glacier equals the rate at which snow and ice melt, the glacier will remain the same size.

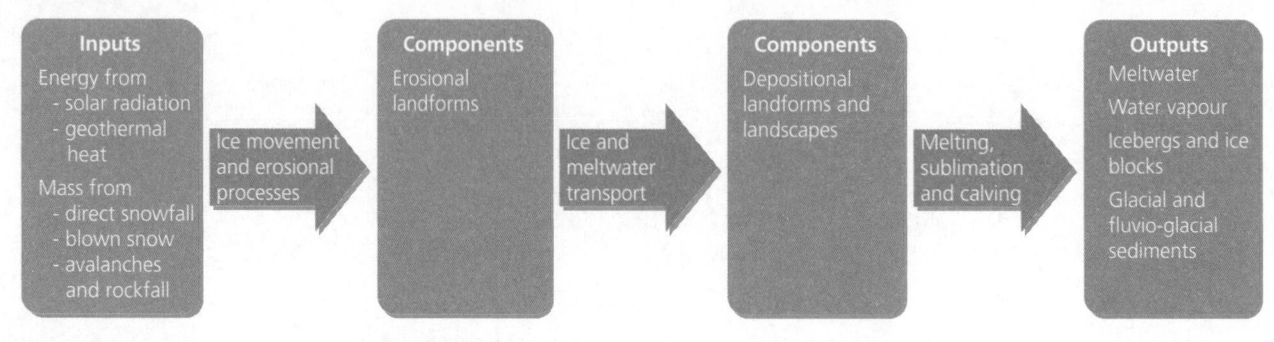

Figure 1.14 A glacier as an open system

If this equilibrium is disturbed, self-regulation will take place to restore the equilibrium, resulting in **dynamic equilibrium**. The system adjusts by a process of **feedback**, which can be either **positive** (an initial change bringing about further change in the same direction) or **negative** (the system is returned to its normal functioning).

Revision activity

Produce a summary diagram including the terms in bold in this section and a definition for each.

Glacier mass balance

The glacial mass balance is also known as the glacial budget (Figure 1.15). It is basically the difference between the inputs and outputs over a year. Inputs include direct snowfall, blown snow and avalanches. Together these inputs are known as **accumulation**.

The inputs are transferred (down valley) by gravity. Mass is lost from the system by melting and evaporation: **ablation**.

Zone of accumulation: upper part of a glacier where input > output.

Zone of ablation: where output > input.

The boundary is the **equilibrium line**.

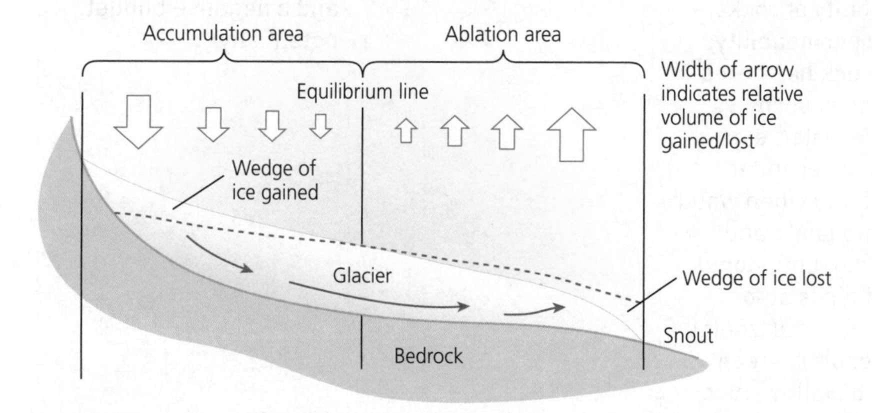

Figure 1.15 **Glacial budget**

Seasonal variations will occur in the glacier mass balance or budget. Accumulation will exceed ablation in winter and ablation will exceed accumulation in summer. Also, because of weather changes from year to year the glacier mass balance varies over time. Climate change will cause changes over a longer timescale.

Now test yourself TESTED

17 What is a glacial budget?
18 How do seasons affect the glacial budget?

Answers on p. 217

Glaciated landscapes are influenced by a range of physical factors

REVISED

Table 1.9 summarises the physical factors influencing glaciated landscapes.

Table 1.9 **Physical factors that influence glaciated landscapes**

Climate	Geology	Latitude and altitude	Relief and aspect
Wind: picks up material and uses it in the processes of erosion, deposition and transportation — **aeolian processes** **Precipitation:** provides the input of snow, sleet and rain. This may have a large seasonal variation **Temperature:** affects input and output to the system — if temperature rises, snow melts and becomes output. In areas of high altitude there may be summer snowmelt and in areas of high latitude temperatures may never rise above zero	**Lithology:** the chemical and physical composition of rocks. This affects the impact of weathering, erosion and mass movement processes, e.g. clay has a weak lithology and erosion is very effective, limestone is vulnerable to chemical weathering because of its calcium carbonate **Structure:** the existence of joints, bedding planes and faults. Also the permeability of rocks — primary permeability when a rock has pores which can absorb and store water, e.g. chalk, and secondary permeability when water seeps into joints and cracks, e.g. limestone. Angle of dip is also influential — horizontal strata result in steep profiles of valley sides, for example	**High latitudes:** e.g. Arctic and Antarctic Circles have cold dry climates, landscapes develop under relatively stable ice sheets **Low latitude** but **high altitude:** landscapes develop under dynamic valley glaciers, e.g. Rocky Mountains	Microclimate can be affected by relief and aspect. **Steep relief** gives more energy to a glacier because of gravity Where an **aspect faces away from the sun** temperatures will remain below zero for longer and so there will be less melting. Glacial budget will be positive. With an **aspect facing the sun** there will be more melting and a negative budget, potentially

Aeolian processes are erosion, transport and deposition by the wind.

Now test yourself

TESTED

19 Define aeolian processes.
20 How does geology influence glaciated landscapes?
21 How does aspect affect a glacial budget?

Answers on p. 217

Exam tip

Remember that physical factors are interconnected; do not view these characteristics in isolation.

There are different types of glacier and glacial movement

REVISED

The formation of glacier ice

Snow initially falls as flakes. Where temperatures are low enough for snow to remain frozen throughout the year, each new snowfall will lead to a building of layers. On settling, the lower layers compact to **firn** or névé (French term for firn). With further compaction air is forced out and a bluish colour is evident. A mass of ice eventually forms a glacier; this may take 30–1,000 years. Glacier ice is not encountered until a depth of about 100 m.

Valley glaciers and ice sheets

Valley glaciers are contained within valleys. They may be outlets from ice sheets or fed from corries. They follow the course of the existing valley as they move downhill.

Ice sheets are large accumulations of ice extending more than 50,000 km^2. There are currently two — Antarctica and Greenland.

Warm-based and cold-based glaciers

Glaciers can be classified as warm- or cold-based.

Warm-based glaciers occur in temperate areas, e.g. in western Norway and southern Iceland. They have the following features:

- They are small (hundreds of metres to a few kilometres in width).
- There is a summer melt.
- Meltwater lubricates the glacier, leading to more movement and consequently more erosion, transportation and deposition.
- All ice in warm-based glaciers is at or above **pressure melting point** because of the warmer atmospheric temperature, the weight of ice and the effect of geothermal heat at the bed.
- There are rapid rates of movement: 20–200 m per year.

Pressure melting point is the temperature at which ice melts when under pressure.

Cold-based glaciers occur in polar areas, e.g. the Arctic and Antarctic.

- They are large: vast ice caps and ice sheets that cover hundreds of square kilometres.
- They occur in areas of low precipitation and little snow. Therefore, there are low levels of accumulation and no melting as the ice stays very cold.
- All ice in cold-based glaciers is below pressure melting point temperature.
- There is very little meltwater and, therefore, slow movement. The glacier is often frozen to the bed of the glacier, meaning less erosion, transport and deposition.
- They have very slow rates of movement: only a few metres per year.

Typical mistake

Note that pressure caused by the ice mass leads to ice melt at temperatures well below freezing.

Basal sliding and internal deformation

Glaciers move because of gravity; however, a few factors influence the movement:

- **gradient:** steeper gradient → more movement
- **thickness of ice:** this will affect basal temperature and pressure melting point
- the **glacial budget:** a positive budget will cause the glacier to advance

There are two types of ice movement. The first involves breaking up into crevasses when solid ice is moving — common in the **upper zone**. The second type involves ice under steady pressure which will 'deform' and move in a more fluid way — common in the **lower zone**. The nature of ice movement can also relate to warm and cold glaciers:

1. **Basal sliding:** common in warm-based glaciers; consists of **slippage**, **creep** and **bed deformation**.
2. **Internal deformation:** common in cold-based glaciers; consists of **intergranular flow** and **laminar flow**.

Extending and compressing flow occurs over ground which varies between steep slopes and less steep slope angles, as shown in Figure 1.16.

Slippage is a circular motion that can cause ice to move away from the back wall of a hollow.

Creep is slow downward movement of loose rock and soil down a gentle slope.

Bed deformation is movement of soft sediment or weak rock beneath a glacier. It is more effective beneath temperate glaciers as the underlying sediment and rock are saturated with water.

Intergranular flow is where individual ice crystals move relative to each other.

Laminar flow is where individual ice crystals move along layers within the glacier.

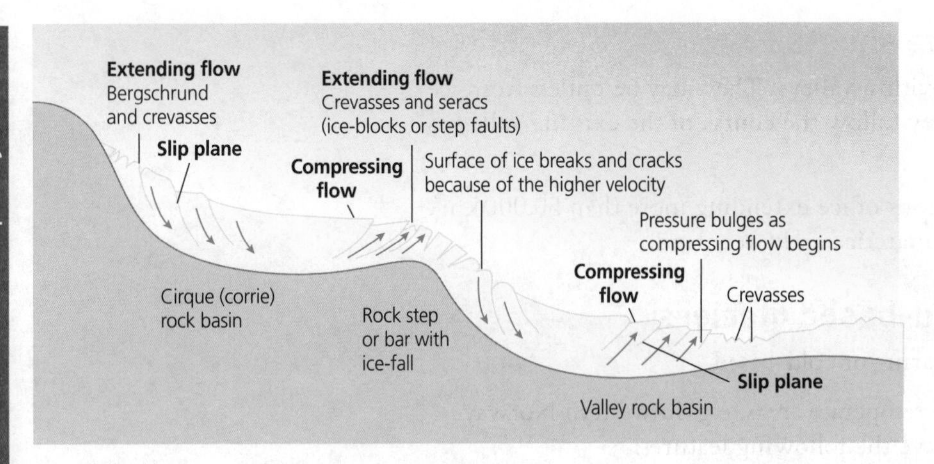

Figure 1.16 Extending and compressing flow

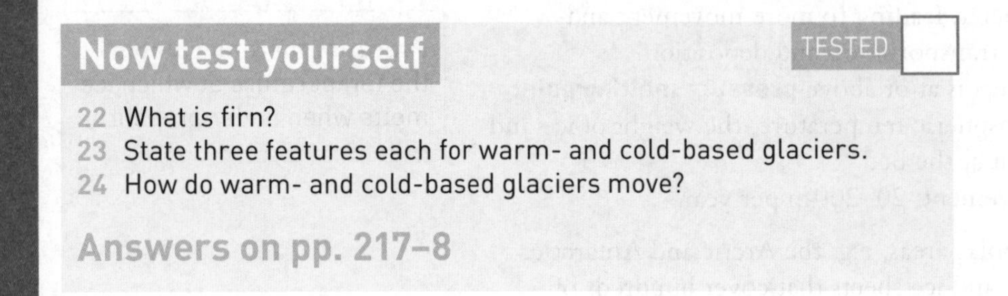

Now test yourself

TESTED

22 What is firn?
23 State three features each for warm- and cold-based glaciers.
24 How do warm- and cold-based glaciers move?

Answers on pp. 217–8

How are glacial landforms developed?

Glacial landforms develop due to a variety of interconnected climatic and geomorphic processes

REVISED

Geomorphic processes

Weathering processes

In glaciated areas, weathering processes can have a significant impact on landscapes. There are three types of weathering processes:

- Physical/mechanical (physical breakdown of rocks into smaller pieces).
- Chemical (involves chemical reactions between weather elements and rock minerals).
- Biological (the action of plants and animals).

These are shown in Table 1.10.

Revision activity

Complete a table describing the different weathering processes in glaciated landscapes. Table 1.12 shows an example of structure. Colour-code the different categories and then in pairs test each other by giving a term and asking for an explanation or give an explanation and ask for the term to which it refers.

Table 1.10 Weathering processes in glaciated landscapes

Mechanical weathering	
Freeze–thaw	Water enters cracks, freezes, and expands by nearly 10%, which exerts a pressure causing the rock to split
Frost shattering	
Pressure release	

Chemical weathering	
Oxidation	
Carbonation	
Solution	
Hydrolysis	
Hydration	
Biological weathering	
Tree roots	
Organic acids	

Mass movement processes

These processes refer to the movement of material (**regolith**) down a slope. They are sub-aerial (above ground) and are dependent on slope angle, particle size, temperature and saturation.

- **Rock fall:** on slopes of 40° or more, material falls to the base of the slope because of gravity.
- **Slides:** may be linear, with movement along a fault or bedding plane, or rotational, with movement along a curved or slip plane.

Glacial erosion processes

- **Abrasion:** material in the glacier is rubbing away at the valley sides and floor. Scratches may be left — **striations**. If the debris is very fine it is called **rock flour**.
- **Plucking:** the glacier freezes onto and into rock outcrops. As the ice moves it pulls away pieces of rock. This mainly occurs at the base of the glacier where jointed rocks have been weakened by freeze–thaw action. Plucking leaves a jagged landscape.

Rates of glacial abrasion are very variable and can be affected by:

- presence of basal debris
- debris size and shape
- relative hardness of particles
- ice thickness
- basal water pressure
- sliding of basal ice
- movement of debris at the base
- removal of fine debris

Nivation is a series of processes operating underneath patches of snow. Freeze–thaw and chemical weathering processes loosen rock, and meltwater removes the debris. The repeat of this over seasons of melting and freezing forms **nivation hollows**.

Glacial transportation

Glaciers carry large amounts of debris from weathering and erosion processes and rock falls from the valley side. Sources of debris for transportation include: rockfall, avalanches, debris flow, aeolian deposits, volcanic eruptions, plucking and abrasion.

Exam tip

Always give as much detail as possible — don't just say rock is weathered, be specific on the types of weathering that relate to the question, e.g. freeze–thaw.

Regolith is a loose layer of rocky material lying over bedrock.

Revision activity

For each of the factors that affect glacial abrasion, indicate if it affects erosion in a positive or negative way, e.g. presence of basal debris will mean more abrasion if there is more debris, so positive.

Typical mistake

The frequency of freeze–thaw cycles is more effective in frost action than in sub-zero temperatures.

Transportation occurs:

- on the surface: supra-glacial
- within the ice: en-glacial
- at the base of the glacier: sub-glacial

Glacial deposition

When the ice melts at the **snout** (the end point of the glacier), material is deposited. Deposition also occurs where the glacier changes between compressing and extending flow. All material deposited during glaciation is known as drift. **Till** (**lodgement till** from advancing ice and **ablation till** from melting ice) is a term used to describe angular, unstratified and unsorted rocks, clay and sand debris deposited by ice. The composition of till reflects the geological conditions over which the ice has travelled. **Outwash** is material deposited by meltwater.

Now test yourself

TESTED

25 What is nivation?
26 Name the different ways in which glaciers transport debris.
27 What is the difference between till and outwash?

Answers on p. 218

The formation of distinctive erosional landforms

Corries, arêtes and pyramidal peaks

Corries (cirques) are armchair-shaped rock basins with a rock lip cut into mountains. Mostly they occur on north- and east-facing slopes where less insolation allows snow to accumulate. Their formation can be sequenced as follows (Figure 1.17).

1. Freeze–thaw weathering above the glacier provides debris which falls onto the top of the glacier.
2. Abrasion occurs at the base of the glacier as it flows forwards and downslope. The depression is deepened.
3. Plucking steepens the back wall and adds debris. The back wall retreats further because of freeze–thaw weathering.
4. Meltwater flows down a deep crevasse (**bergschrund**) which opens up between the glacier and the back wall. This water aids movement (as it lubricates the base of the glacier). Abrasion continues and, as the meltwater freezes, plucking and freeze–thaw processes also occur.
5. The glacier moves in a rotational pattern and continued erosion deepens the basin.
6. At the outlet of the basin ice movement is upwards, there is less erosion and a lip forms.

Exam tip

In explanations always follow a clear sequence of events and refer to named processes rather than just saying 'by erosion'. Make clear the link between processes and landforms.

A **bergschrund** is a type of crevasse formed near the back wall of a corrie. It forms as the glacier starts to move away.

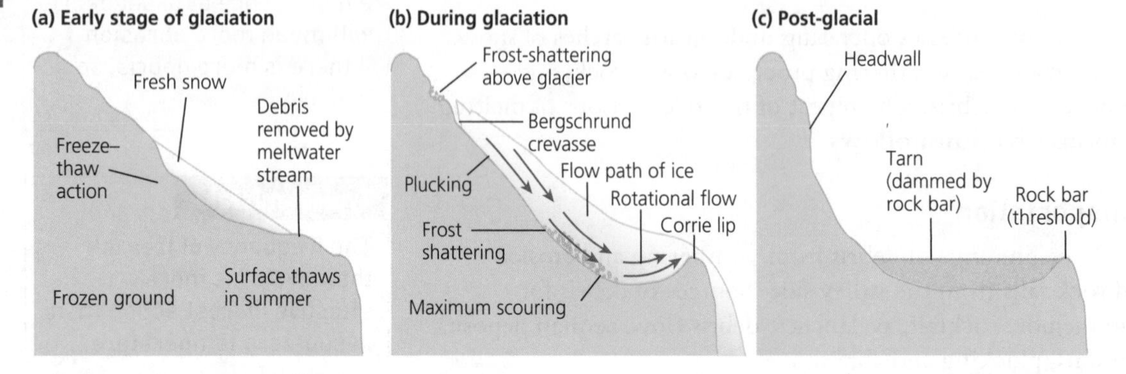

Figure 1.17 The formation of a corrie

An arête is formed where two corries lie back-to-back alongside each other. If more than two corries develop on a mountain, the remaining central mass forms a pyramidal peak.

Examples

Corrie: Red Tarn Corrie, Lake District
Arête: Striding Edge above Red Tarn, Lake District
Pyramidal peak: Machhapuchhre, Nepal

Typical mistake

Note that many glacial features are formed when landscapes are buried beneath ice sheets, e.g. in Antarctica today.

Glacial troughs, hanging valleys, truncated spurs and roches moutonnées

Glaciers flow down pre-existing river valleys. Because of their power, they deepen these valleys and change the V shape to a U shape. These U-shaped valleys are straight, wide-based and steep-sided: this is a **glacial trough**. A **hanging valley** is the result of differential erosion between a main glacier and the valley of a tributary. The tributary is eroded at a slower rate so that when the ice melts the tributary valley is left 'hanging' high above the main valley. Often, single or multiple waterfalls join the tributary to the main post-glacial river. These features are explained in Figures 1.18 and 1.19.

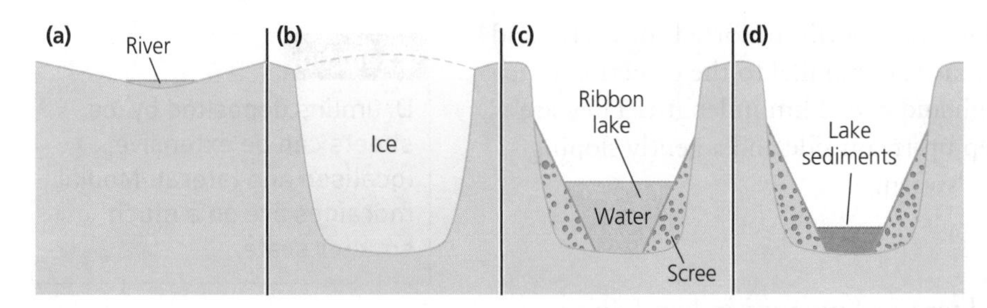

Figure 1.18 The formation of a glaciated valley

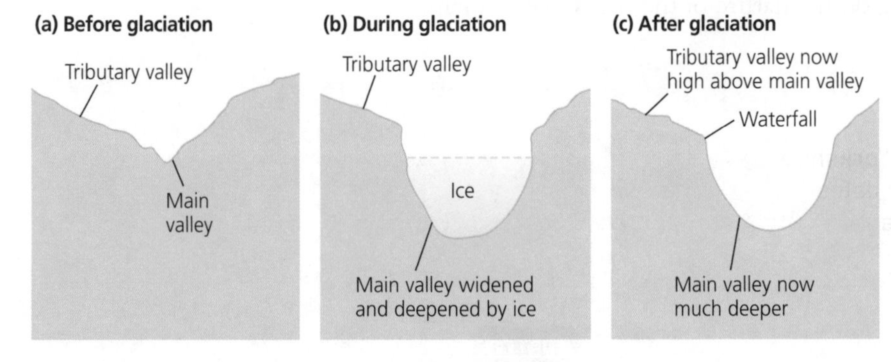

Figure 1.19 The formation of a hanging valley

Exam tip

Glacial landforms are complex and the result of numerous glacial periods over the past 1 million years.

The ice, meltwater and sub-glacial debris combined have huge erosive power. Several further landforms are associated with glacial troughs:

- Areas of land protruding from the river valley side (spurs) are removed by the glacier, forming **truncated spurs**.
- Areas of resistant rock on the valley floor are not completely removed and are left as **roches moutonnées**. They have a smooth up-valley side created by abrasion and a jagged down-valley side due to the action of plucking.

Exam tip

Landforms are small-scale features, while landscapes refer to the way features interconnect to form a landscape.

Striations

Striations are scratches or grooves made by debris embedded in the base of a glacier.

The formation of distinctive depositional landforms

Moraines

Moraines are landforms created when the debris carried by a glacier is deposited. There are several types of moraine:

- **Lateral:** derived from frost shattering of the valley sides, carried at the edge of a glacier; on melting, a side embankment is formed.
- **Medial:** the merging of two lateral moraines.
- **Terminal:** a high mound extending across the valley to mark the maximum advance of the ice sheet.
- **Recessional:** mark an interruption in the retreat of the ice.
- **Push:** form if the climate deteriorates and the ice advances.

Erratics

Erratics are individual pieces of rock picked up and carried by the ice (often over many kilometres) to be deposited in areas of a completely different geology.

Drumlins

Drumlins are smooth, elongated mounds of till (unsorted rock, clay and sand deposited by ice). The long axis runs parallel to the direction of ice movement. They can be 50 m high and over 1 km in length. They are smoothed by abrasion with a steep upstream side and a gently sloping downside. In a group they form a 'swarm'.

Exam tip

Drumlins deposited by ice sheets can be extensive, localised and lateral. Medial moraines are on a much smaller scale.

Till sheets

Till sheets are often found behind terminal moraine in low-lying areas. They are wide areas of flat relief where there is a covering of glacial till (sand and gravel). They are significant because of their extent. They are variable in composition depending on the nature of the rocks over which the ice has moved.

Examples

Drumlins: Risebrigg Hill, North Yorkshire
Erratic: Ingleborough, Yorkshire dales
Moraines: Meade Glacier, Alaska
Till sheet: Minnesota, USA

Now test yourself

TESTED

28 Name two erosion processes that contribute to the creation of a corrie.

29 What are:
(a) a drumlin
(b) lateral moraine?

Answers on p. 218

Glacial landforms are inter-related and together make up characteristic landscapes

REVISED

Case study: One landscape associated with the action of valley glaciers

Online (see p. 3 for details) you will find a sketch of Snowdonia, illustrating a landscape associated with the action of a valley glacier.

Revision activity

Research a photograph or sketch of a landscape associated with the action of valley glaciers. Summarise your case study revision notes by annotating the resource, as in the sketch of Snowdonia.

One landscape associated with the action of ice sheets

Online (see p. 3 for details) you will find a case study on Minnesota, USA, illustrating a landscape associated with the action of ice sheets.

How do glacial landforms evolve over time as climate changes?

Glacio-fluvial landforms exist as a result of climate change

REVISED

Glacio-fluvial landforms are produced by meltwater from glaciers during deglaciation. The resulting streams often flow at high velocity. A loss of energy due to a decrease in discharge leads to deposition, although landforms of erosion and deposition are characteristic of this type of landscape.

Post-glacial climate change and its effect on geomorphic processes

When global temperatures rise, glacial periods end and interglacials lasting 10,000–15,000 years follow. Glacio-fluvial rivers deposit outwash which has the following features:

- small material carried by meltwater streams with less energy than ice
- smooth and rounded particles
- sorted, with larger material further up the valley
- stratified vertically with distinct seasonal and annual layers

Exam tip

Remember to draw knowledge on climate change from other parts of the course.

Glacio-fluvial landforms

Figure 1.20 shows some of the features of glacio-fluvial landscapes.

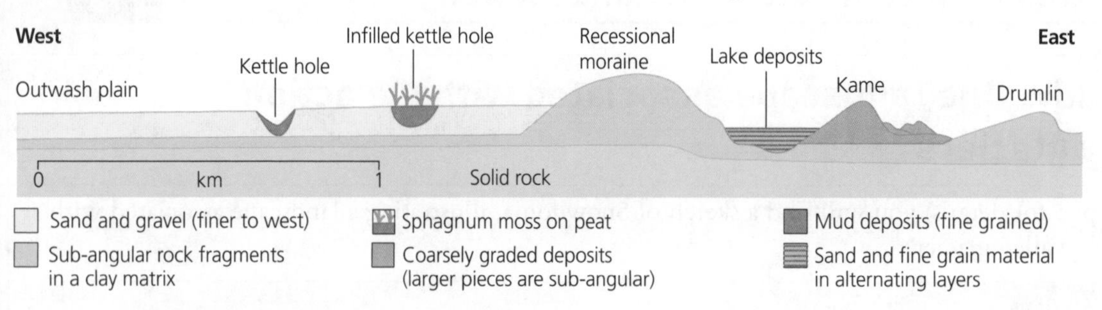

Figure 1.20 Features of glacio-fluvial landscapes

Kames

Kames are undulating, winding mounds of unevenly deposited sand and gravel. Kame terraces are flat areas formed along the side of valleys. They follow the direction of ice advance.

Eskers

Eskers are very long, narrow ridges of sorted, stratified, coarse sand and gravel. Deltaic deposits are left when meltwater flows into a lake trapped by moraine deposits (Figure 1.21).

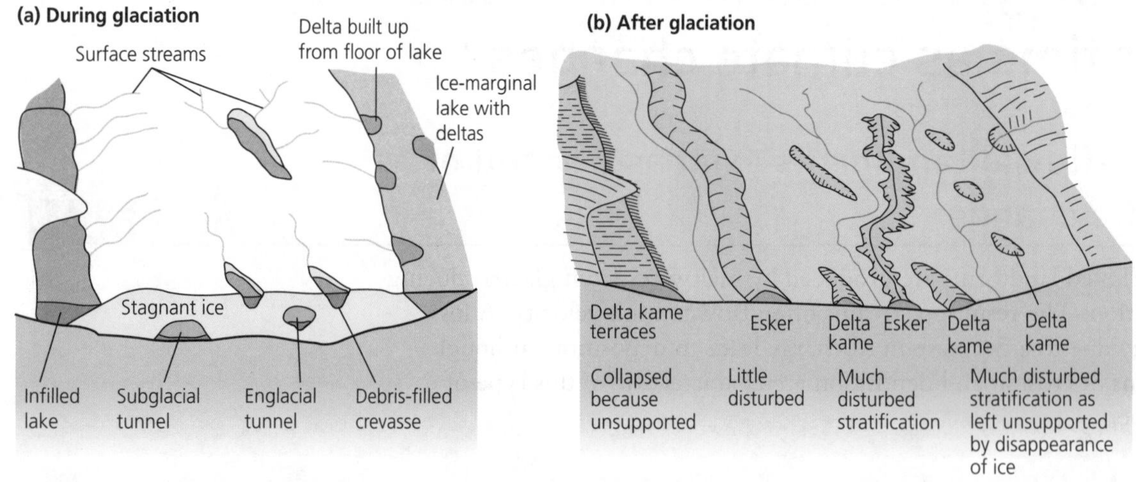

Figure 1.21 Formation of kames and eskers

Outwash plains (sandur)

Sandurs are deposits by meltwater streams running out from the snout (end) of the glacier. Composition is of coarse material, which is found near to the glacier, and finer clay, which is carried across the plain before being deposited.

Meltwater streams that cross the outwash plain are braided (they divide as the channels become choked with material). There are also often a series of small depressions called kettle holes, formed when blocks of ice that have been washed onto the plain melt and leave a gap in the sediments.

An example of a glacio-fluvial landscape and outwash plains is Skeidarár Sandur, Iceland.

Typical mistake

It is important to know how glacio-fluvial and ice deposits differ. Glacio-fluvial deposits tend to be layered and sorted. Ice deposits are unstructured.

Modifications of landforms due to changes in climate

The appearance of glacio-fluvial landscapes is modified by repeated advance and retreat of ice and by weathering, erosion and vegetation colonisation in post-glacial times. Rising temperatures lead to more melting and expanses of outwash; also vegetation such as mosses, lichens and then grasses will colonise landscapes.

Now test yourself TESTED

30 What is an esker?
31 How do glacio-fluvial deposits differ from ice deposits?

Answers on p. 218

Periglacial landforms exist as a result of climate change

REVISED

Periglacial environments are areas with:

- **permafrost**
- seasonal temperature variations with a brief 'summer' above 0°C
- freeze–thaw cycles

Permafrost is permanently frozen soil and regolith.

Periglacial environments are found in high latitudes, e.g. Alaska; continental interiors, e.g. Siberia; and high mountains at lower latitudes, e.g. the Andes.

Typical mistake

Note that permafrost has a brief summer thaw of the top layer — not all the ground is permanently frozen.

Climate changes and the effect on geomorphic processes

As a result of seasonal fluctuations in temperature around freezing, freeze–thaw weathering is a dominant process in periglacial environments. The development of ground ice and the process of frost heave are also important.

Periglacial landforms

Patterned ground

Patterned ground is a landform reflecting the repeated cycles of freezing and thawing of the active layer. Rock particles are distributed in a system of polygons and circles. The process of frost heave (expansion of the volume of soil as ice crystals form) pushes larger stones to the surface and, because of the camber, stones move sideways (see Figure 1.22 on p. 40).

Ice wedges

Ice wedges are narrow, frost-formed cracks in the upper layers of the ground which fill with ice. They can be up to 10 m in depth.

Pingos

A pingo is a dome-shaped, ice-cored mound of earth. There are two theories of the formation of pingos (Table 1.11).

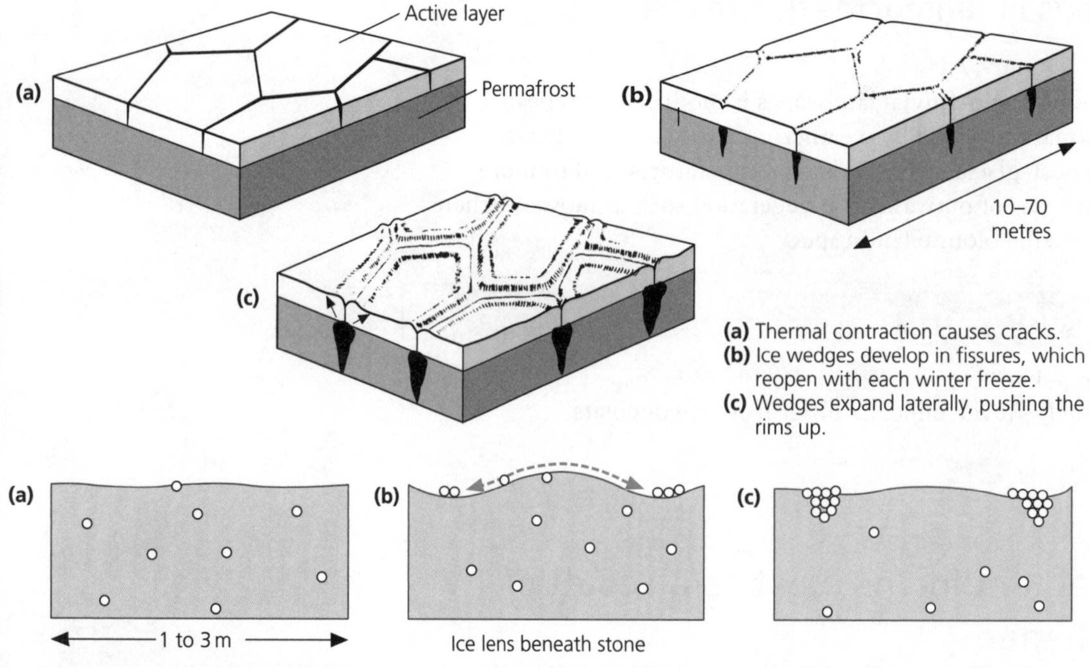

Figure 1.22 Patterned ground

Table 1.11 The formation of pingos

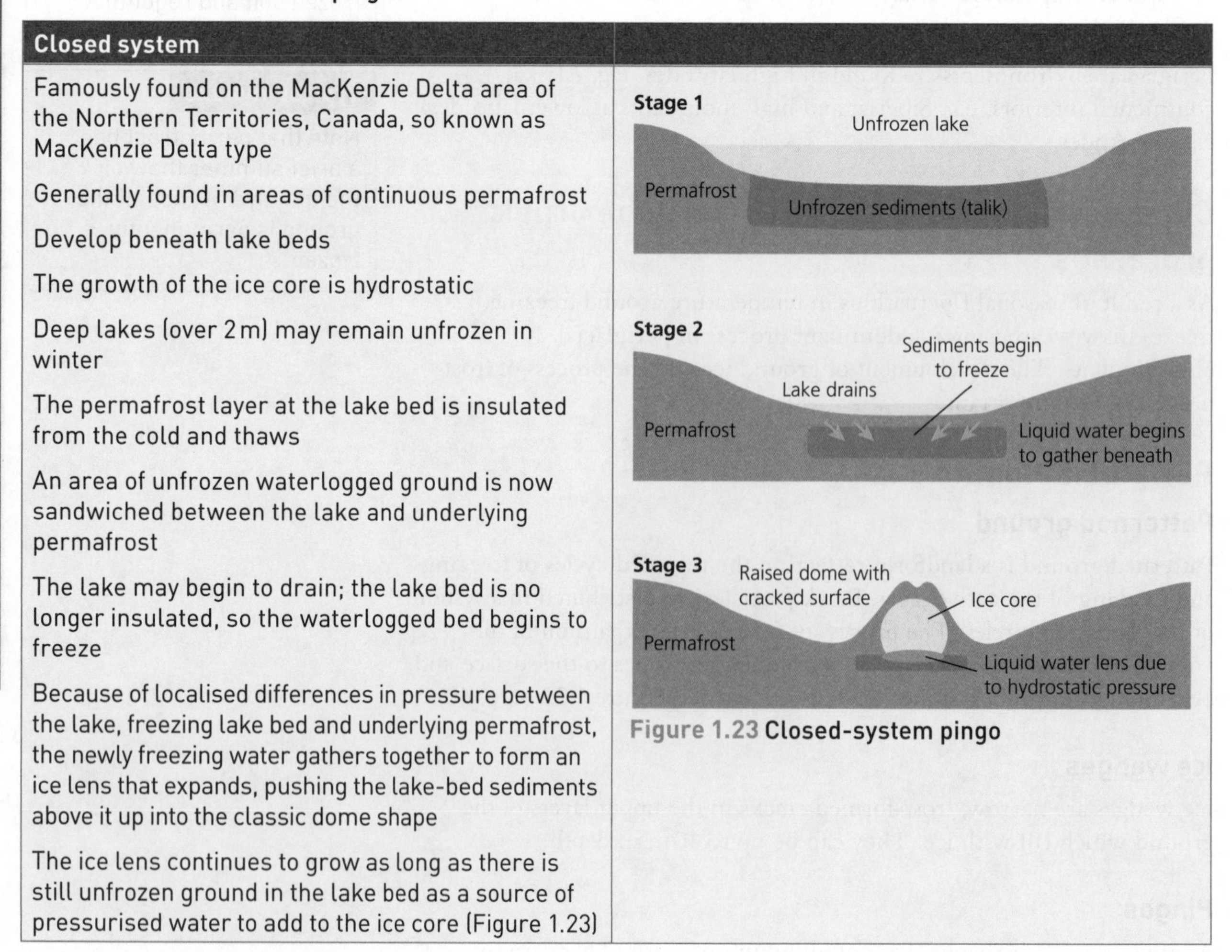

Closed system	
Famously found on the MacKenzie Delta area of the Northern Territories, Canada, so known as MacKenzie Delta type Generally found in areas of continuous permafrost Develop beneath lake beds The growth of the ice core is hydrostatic Deep lakes (over 2 m) may remain unfrozen in winter The permafrost layer at the lake bed is insulated from the cold and thaws An area of unfrozen waterlogged ground is now sandwiched between the lake and underlying permafrost The lake may begin to drain; the lake bed is no longer insulated, so the waterlogged bed begins to freeze Because of localised differences in pressure between the lake, freezing lake bed and underlying permafrost, the newly freezing water gathers together to form an ice lens that expands, pushing the lake-bed sediments above it up into the classic dome shape The ice lens continues to grow as long as there is still unfrozen ground in the lake bed as a source of pressurised water to add to the ice core (Figure 1.23)	**Figure 1.23** Closed-system pingo

Open system	
Common in Greenland and Alaska, so known as East Greenland type Generally found in areas of discontinuous permafrost Found in valley bottoms The growth of the ice core is hydraulic Water is able to seep into the upper layers of the ground and flows from higher surrounding areas under artesian pressure Water accumulates in flat, low-lying areas between the upper layers of permafrost or soil and frozen ground beneath the water, and then freezes The freezing ice core expands, thus doming the overlying layers into the classic pingo shape They can grow as pressurised water continues to flow in from their surroundings (Figure 1.24)	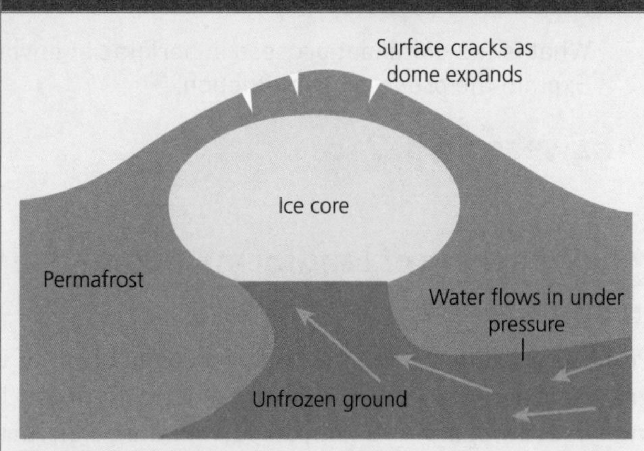 **Figure 1.24 Open-system pingo**

Blockfields

Freeze–thaw action produces large amounts of scree which forms scree slopes. In flat areas expansive areas of angular boulders form blockfields.

Solifluction

Summer melt of water in the upper layers of permafrost leads to large amounts of water which cannot drain away because of the permafrost. The lubrication means that soil is moved even on the most gentle of slopes. Figure 1.25 shows how periglacial landforms combine to form a characteristic periglacial landscape.

Exam tip

Remember that a key feature of the Landscape systems unit is understanding how landforms combine to form characteristic landscapes.

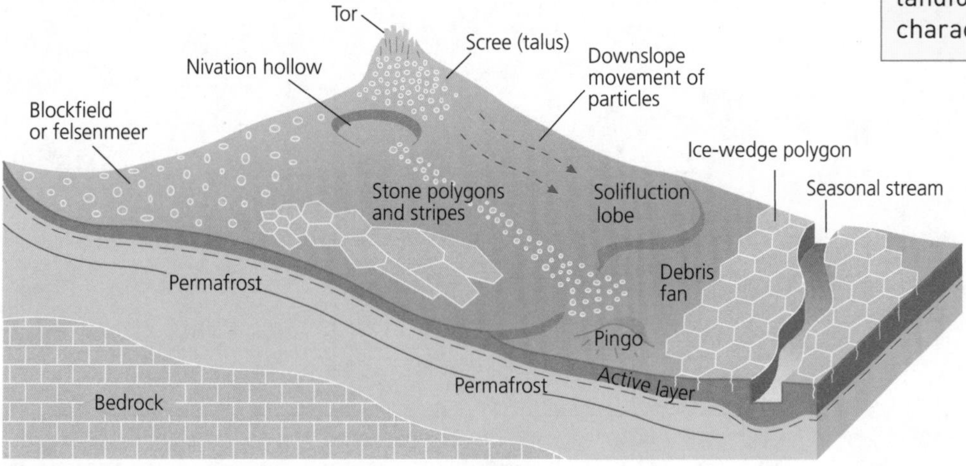

Figure 1.25 Features of periglacial landscapes

Examples

Patterned ground: Barrow, Alaska
Pingo: northern Canada
Blockfields: Scafell Ranges, Lake District
Solifluction sheet: Ogilvie Mountains, Canada

Solifluction is the slow flow of saturated regolith down a gradient which can be very gentle.

Now test yourself

TESTED

32 What is the dominant process in periglacial environments?
33 Explain the process of solifluction.

Answers on p. 218

Exam tip

Read the specific command terms of exam questions carefully. Rarely will you be asked to just 'describe' a landform.

Modifications of landforms due to changes in climate

Periglacial landforms are modified by colonisation of vegetation, e.g. grasses. Pingos also collapse when the ice core thaws, leaving a feature known as an ognip — a rampart surrounding a circular depression.

Revision activity

Using the table headings below, a complete checklist of the characteristic landforms in each of the summary landscapes outlined above.

Table 1.12 Landforms of glaciated landscapes

Erosional landforms in glaciated landscapes	Depositional landforms in glaciated landscapes	Landforms of glacio-fluvial landscapes	Landforms of periglacial landscapes
Name of landform:			
Dominant process(es):			
Examples:			

How does human activity cause change within glaciated and periglacial landscape systems?

Human activity causes change within periglacial landscape systems

REVISED

Case study of **one** periglacial landscape that is being used by people

Revision activity

Select a suitable method of presentation, e.g. star diagram or annotated map, and summarise the case study covered in class for a periglacial landscape that is being used by people. Focus on the following:

- the human activity taking place and the reasons for it
- the impacts on processes and flows
- the effect of these impacts in changing periglacial landforms
- the consequence of these changes on the landscape

Human activity causes change within glaciated landscape systems

REVISED

Case study of **one** glaciated landscape that is being used by people

Revision activity

Select a suitable method of presentation and summarise the case study covered in class for a glaciated landscape that is being used by people. Focus on the following:

- the human activity taking place and the reasons for it
- the impacts on processes and flows
- the effect of these impacts in changing glacial landforms
- the consequence of these changes on the landscape

Summary

- The systems approach can be applied to the study of glaciers, with glacial budget and mass balance being key concepts.
- A number of physical factors influence glaciated landscapes.
- There are different types of glacier and glacier movement.
- Classic glaciated landscapes are formed, which include a range of erosional and depositional landforms.
- Glacial landforms change over time as climate changes.
- Glacio-fluvial and periglacial landforms exist as a result of climate change.
- Human activity in periglacial and glaciated landscapes causes change with impacts and consequences.

Exam practice

6 Using evidence from the case study of a landscape associated with the action of a valley glacier (Snowdonia, p. 37), explain the impact of valley glaciers on the landscape. [3]

7 Suggest why weathering processes are more effective in glacial landscapes where temperatures can rise above 0°C. [4]

8 Explain the formation of a corrie. [8]

9 Explain the concept of a glacial budget. [8]

10 Using the data in Table 1.13:

(a) Calculate the interquartile range (IQR). [2]

(b) A fieldwork study of the relationship between time and Arctic sea ice extent gave a Spearman's rank correlation result of R_s–0.582 for 20 paired values. Using Table 1.14, state the null hypothesis and interpret this result. [4]

Table 1.13 **Snowfall at selected weather stations in Alaska**

Weather station	1	2	3	4	5	6	7	8	9	10	11
Average annual snowfall (cm)	365	665	190	120	160	225	90	170	200	100	300

Table 1.14

Degrees of freedom	0.05 significance level	0.01 significance level
19	±0.388	±0.549
20	0.377	0.534
21	0.368	0.521

Answers and quick quiz 1B online

ONLINE

Option C Dryland landscapes

How can dryland landscapes be viewed as systems?

Dryland landscapes can be viewed as systems

REVISED

Drylands are regions where average annual evapotranspiration is significantly higher than precipitation. Arid or semi-arid areas cover approximately 40% of the land surface. They exist in mid-, low- and high-latitude areas (Figure 1.26).

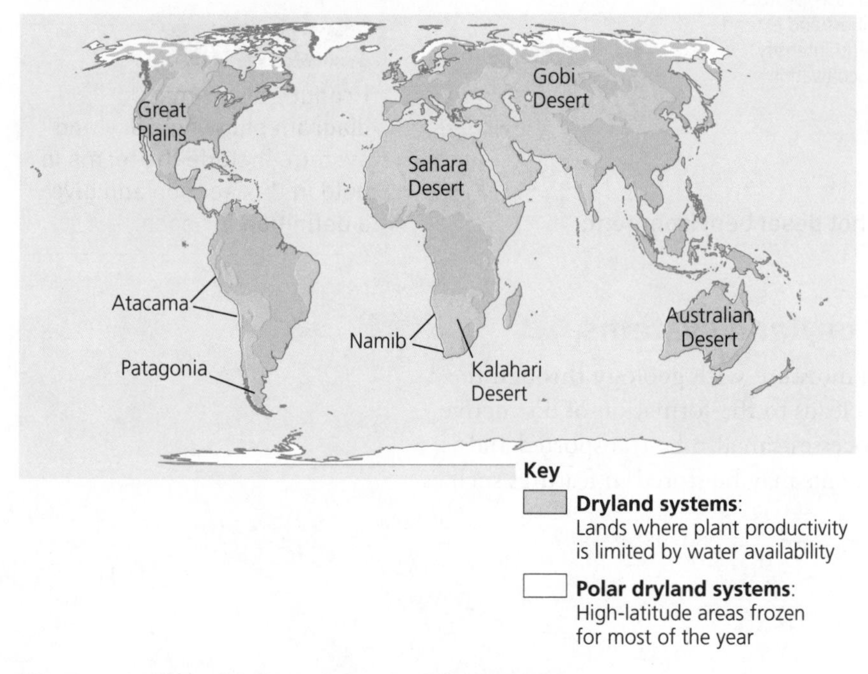

Figure 1.26 Distribution of the world's drylands

Revision activity

Make summary bullet-point notes or annotate a copy of Figure 1.26 to describe and account for the distribution of the world's drylands.

The components of dryland systems

The system approach is a way of analysing the relationships within a unit. It consists of several components (stores) and processes (links) that are connected and can be represented in a flow diagram.

A dryland system is an **open system**, meaning that energy and matter can cross the boundary of the system to the surrounding environment. It has:

- **inputs:** solar radiation and precipitation
- **outputs:** heat (long-wave radiation from the surface), evapotranspiration and stream flow

The combination of these factors forms distinctive landscapes which are made up of a range of erosional and depositional landforms created by natural geomorphic processes and reflecting human activity.

When the inputs and outputs of a system are equal it is in a state of **equilibrium**. However, dryland systems are dynamic (constantly changing) places and the equilibrium is often disturbed, resulting in **dynamic equilibrium**.

Change occurs to upset the balance of the system through human activity, for example. The system adjusts by a process of **feedback**, which can be either **positive** (an initial change bringing about further change in the same direction) or **negative** (the system is returned to its normal functioning).

An example of positive feedback in a hot desert environment is shown in Figure 1.27.

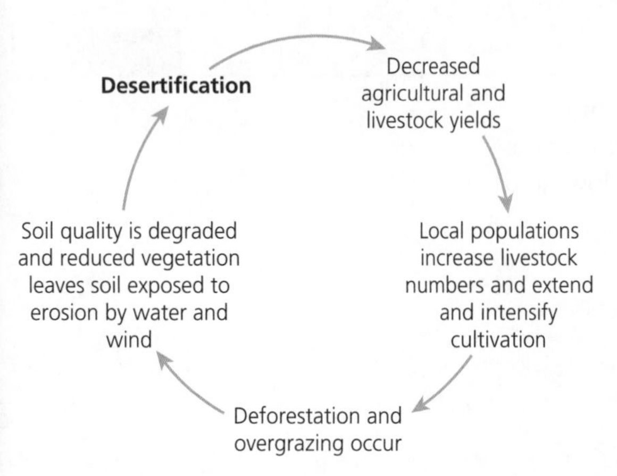

Figure 1.27 Positive feedback in a hot desert environment: desertification

Desertification is a reduction in agricultural productivity due to overexploitation of resources and to natural processes such as drought.

Revision activity

Produce a summary diagram showing a dryland system. Include the terms in bold in this section and give a definition for each.

Flows of energy through dryland systems

The interaction of temperature and moisture with geology through the processes of weathering and erosion leads to the formation of distinctive landforms. Materials from these processes can also be transported and deposited by wind and rivers. Sediments may be stored in features such as screes and dunes (Figure 1.28).

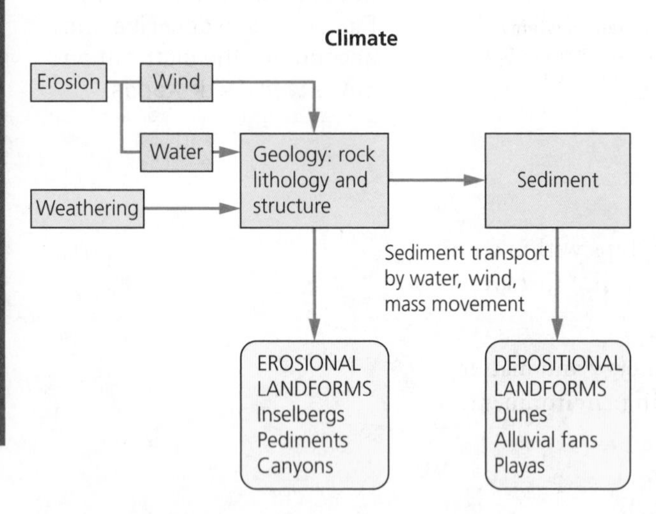

Figure 1.28 Flows in a dryland system

Aridity index and UNEP

Calculations of aridity are based on **potential evapotranspiration** (PET). PET takes account of atmospheric humidity, solar radiation and wind. In drylands annual PET always exceeds annual precipitation. Moisture that could be lost through evapotranspiration is greater than the amount of moisture available.

Potential evapotranspiration (PET) is the amount of evaporation that would occur if sufficient water sources were available.

The United Nations Environment Programme (UNEP) aridity index provides a means of measuring aridity:

$$AI = \frac{P}{PET}$$

where AI is aridity index; P is mean annual precipitation; and PET is mean annual potential evapotranspiration.

Typical mistake

Note that aridity is more than just low rainfall; it is also related to dryness in the climate, i.e. humidity levels.

Now test yourself

TESTED

34 What factors are taken into account in PET?
35 What is an aridity index and how is it measured?

Answers on p. 218

Dryland landscape systems are influenced by a range of physical factors

REVISED

Table 1.15 summarises the physical factors influencing dryland landscapes.

Table 1.15 Physical factors that influence dryland landscapes

Climate	Low annual precipitation Potential evapotranspiration much higher than precipitation Wind is an active agent of landscape change
Geology	Lithology: chemical composition and structure — physical characteristics, e.g. joints, determine the porosity (degree to which rock holds water) and permeability (degree to which water is absorbed through joints and cracks in the rock)
Latitude and altitude	Latitude influences drylands. The atmospheric circulation of the low latitudes influences deserts such as the Sahara through the forces of the Hadley cell (Figure 1.29) Latitude also influences the amount and intensity of incident solar radiation — average temperatures decrease polewards Cold ocean currents influence drylands between latitudes 15° and 30° (Figure 1.30) Altitude has a more local influence on aridity. Temperatures fall with altitude; cloudless skies lead to large temperature ranges, low humidity and small amounts of precipitation
Relief and aspect	Leeward slopes of mountains can be relatively dry as they lie in the **rain shadow** Southerly aspects lead to warmer temperatures and higher rates of evapotranspiration, leading to more dryness
Availability of sediment	Sediment is material of varying size derived from weathering and erosion processes. Rivers, wind and mass movement processes transport and deposit sediment in drylands In some drylands sediment is scarce and desert pavements are a feature In drylands where there is a continuous supply of sediment, possibly from more active weathering processes, **alluvial fans**, **bajadas** and **sand seas** are features

A **rain shadow** is an area of below-average rainfall situated in the lee of an upland.

An **alluvial fan** is a cone of sediment deposited by a river where it leaves a steep upland course and flows onto low land.

A **bajada** is a series of alluvial fans that merge to form a continuous spread of sediment.

A **sand sea** is a vast expanse of sand dunes.

Exam tip

Remember that physical factors are interconnected; do not view these characteristics in isolation.

Exam tip

Rainfall effectiveness is important in hot arid areas. There is rainfall but it quickly evaporates before it can be effective.

Now test yourself

TESTED

36 Define the rain shadow area.
37 How does altitude affect dryland landscapes?
38 How does aspect affect dryland landscapes?

Answers on p. XX

Cause and effect explanation of Hadley cells

At the equator there is a large amount of solar radiation because the sun is directly overhead → air in contact with the land is heated → it rises and cools, water vapour condenses → precipitation forms → the rising air is replaced with air from the north and south, creating low-pressure areas — the inter-tropical convergence zone (ITCZ) → the rising air tracks polewards and at 20°–30° north and south this now cooler air descends → it warms as it descends, and it expands, resulting in little cloud, clear skies and aridity in these latitudes. The cells of circulating air are known as the **Hadley cells** (Figure 1.29).

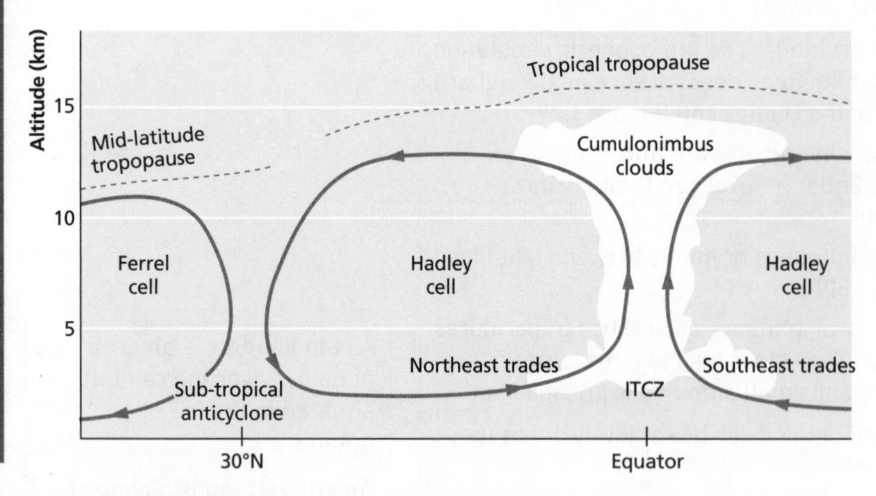

Figure 1.29 Hadley cell

Revision activity

It is essential to practise cause and effect sequences of explanations for examination answers. Hadley cells are explained in this section — write out a cause and effect explanation for (a) the influence of cold ocean currents and (b) the rain shadow effect.

The influence of cold ocean currents

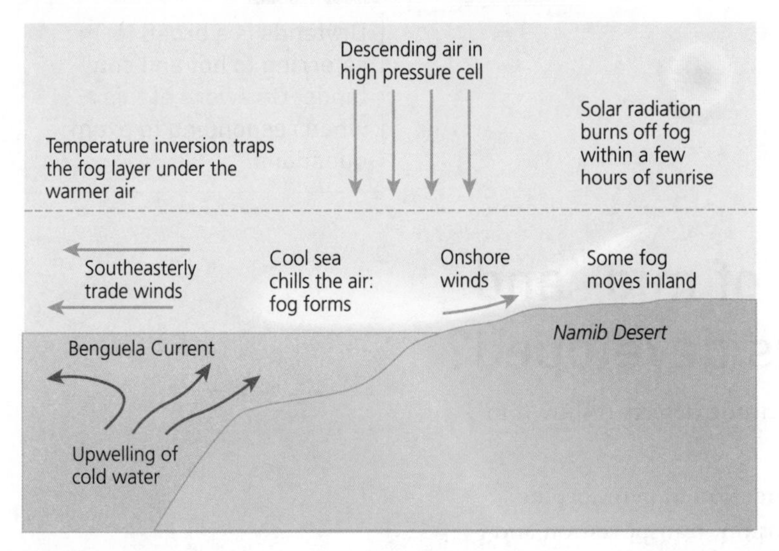

Figure 1.30 The influence of cold ocean currents on the Namib Desert

Typical mistake

Temperature changes are influenced by changes in pressure as well as radiation. Descending air is compressed and warms; rising air expands and cools.

Exam tip

A combination of physical factors often causes drylands to form. Be specific in relating physical factors to any named example.

There are different types of dryland

REVISED

There are three main types of dryland landscapes: polar drylands, mid- and low-latitude deserts, and semi-arid environments (Figure 1.31).

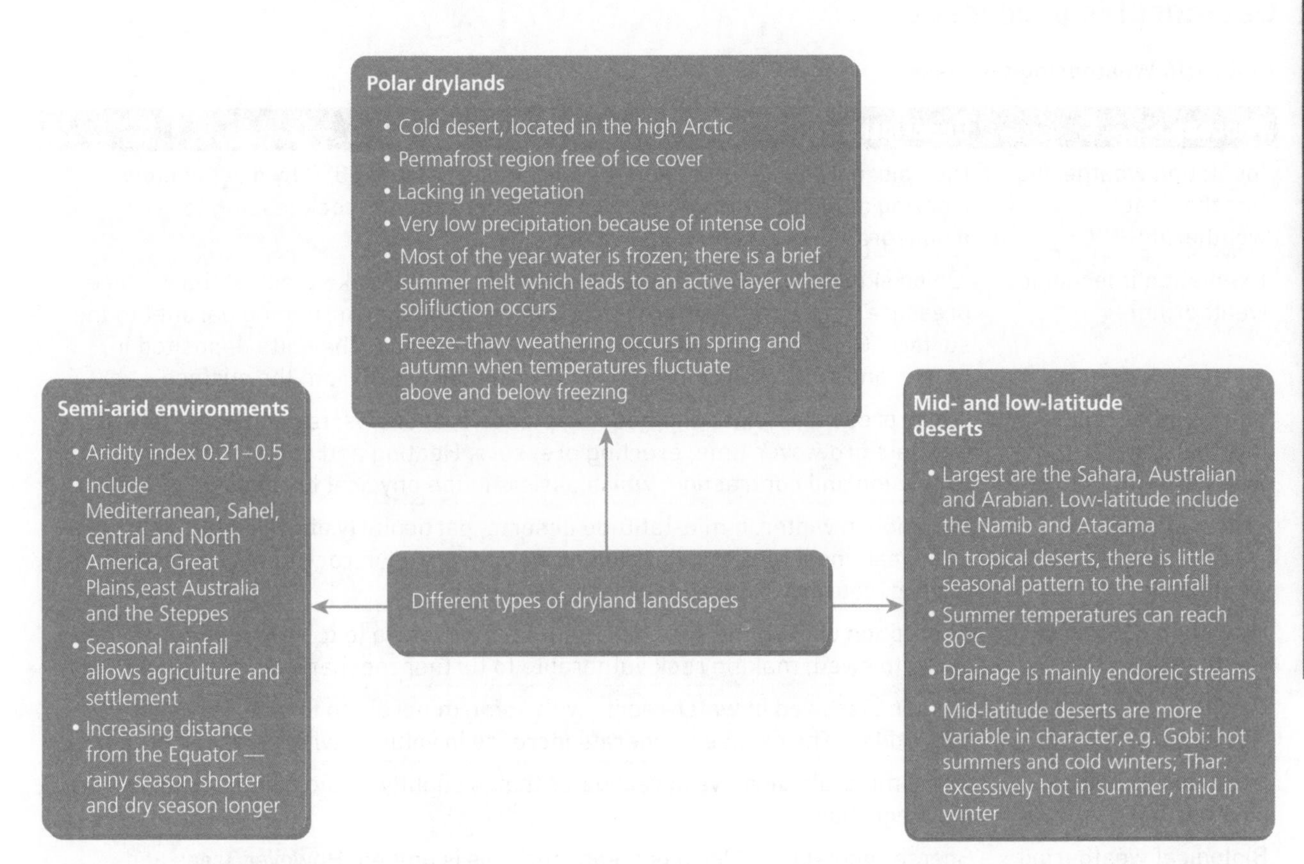

Figure 1.31 Different types of dryland

Now test yourself

TESTED

39 What is tundra?
40 Explain the process of solifluction.
41 State **three** types of dryland.

Answers on p. 218

Exam tip

Drylands is a broad term referring to hot and cold lands. Be aware of this when responding to exam questions.

How are landforms of mid- and low-latitude deserts developed?

The location of the major mid- and low-latitude deserts is shown in Figure 1.26, p. 45.

Desert landforms develop through the interaction of geomorphic processes (weathering, mass movement, erosion, transportation and deposition) with rock formations.

Dryland landforms develop due to a variety of interconnected climatic and geomorphic processes

REVISED

Geomorphic processes

Table 1.16 Weathering processes

Process	Description
Insolation weathering (mechanical weathering)	High diurnal temperature ranges (variations of from 80°C by day to below freezing at night) cause expansion and contraction in rock leading to disintegration over a long period of time
Exfoliation (mechanical weathering)	'Onion skin weathering'. As weathering and erosion take place at the surface, pressure is released from rocks at depth. Cracks form running parallel to the surface. Capillary action brings salts to the surface. The salts deposited in cracks and enhanced chemical weathering peel rock from the surface
Salt weathering (mechanical weathering)	Water present in joints and bedding planes evaporates, leaving salts behind. Crystals grow over time, exerting pressure. Heating and cooling lead to expansion and contraction, which assists in the physical breakdown of rock
Freeze–thaw weathering (mechanical weathering)	Common in winter in mid-latitude deserts, particularly above 1,500 m altitude. Significant heat loss at night causes water trapped in rock joints to freeze and expand, causing the gradual break-up of rocks
Hydration (chemical weathering)	Absorption of even the smallest amount of moisture (e.g. dew and fog) causes rocks to swell, making rock vulnerable to further mechanical breakdown
Oxidation (chemical weathering)	Oxygen dissolved in water reacts with some minerals to form oxides and hydroxides. The oxidised minerals increase in volume, which weakens the rock
Solution (chemical weathering)	Some minerals dissolve in rainwater that is slightly acidic and this causes rock disintegration
Biological weathering	Sparse vegetation in deserts means that this is limited. However, trees and shrubs have long root systems that can widen joints and dislodge small particles
Block and granular disintegration	The processes above can lead to the breakdown of rock into large blocks where bedding planes and joints are prominent. As mechanical and chemical processes take effect, **block disintegration** occurs. Where individual grains are broken away from rock surfaces by the effects of thermal extraction and contraction or freeze–thaw action of moisture, **granular disintegration** occurs

Revision activity

In pairs and using Table 1.16, test each other on key weathering processes. Either give the term and ask for an explanation or give an explanation and ask for the term to which it refers.

Typical mistake

Do not overlook weathering processes in the explanation of landforms in hot drylands. There is enough moisture available in a hot desert to allow a range of weathering processes to take place. Even small amounts of dew are enough.

Mass movement

- **Debris flow:** large quantities of rock fragments, mud and soil move downslope at speed after heavy rainfall and where vegetation is sparse.
- **Rockfalls and rock slides:** where resistant rock such as sandstone rests on a weaker rock such as shale, undercutting at the base of the slope can be destabilising.

Fluvial and aeolian processes

Figure 1.32 summarises the main fluvial and aeolian processes.

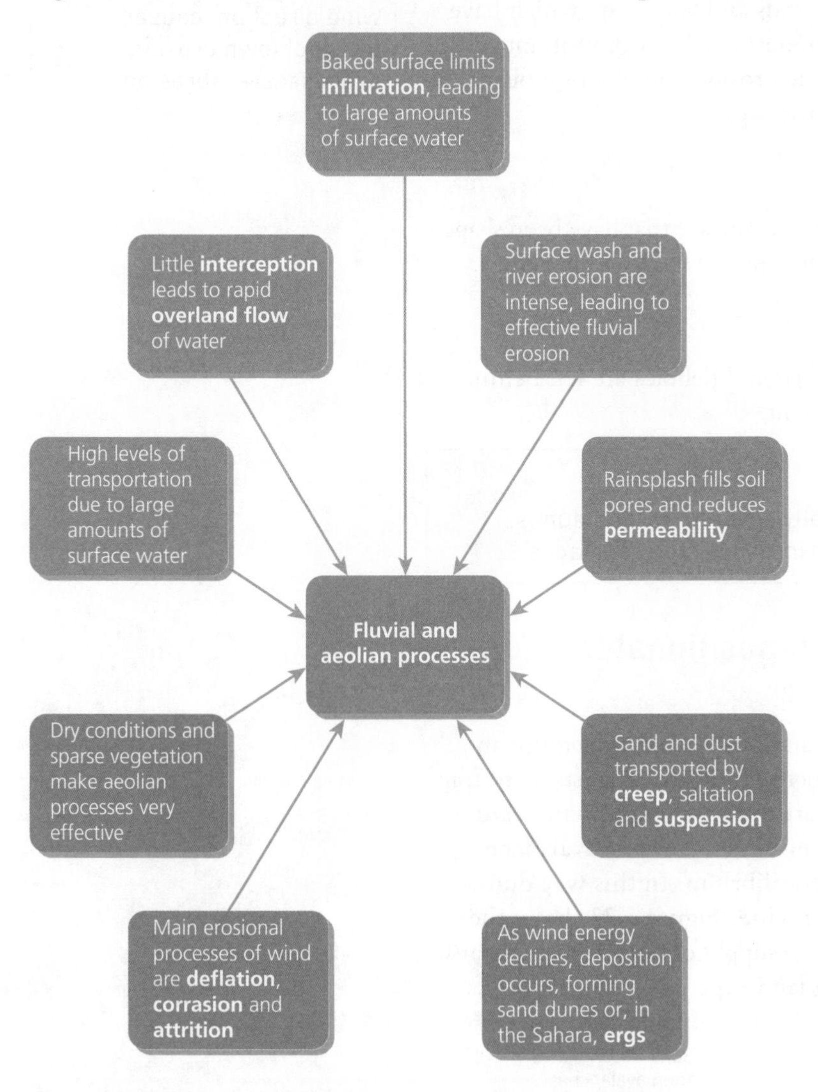

Figure 1.32 Fluvial and aeolian processes in drylands

Now test yourself

TESTED

42 What is an ephemeral stream?

43 Name the **three** ways in which sand is transported.

Answers on p. 218

The formation of distinctive erosional landforms

Wadis

Wadis are steep-sided, wide-bottomed, gorge-like valleys formed by fluvial erosion. They are rarely filled with water and have a build-up of sediment on the valley floor. They are either permanently dry or have ephemeral streams that are fast-flowing and the result of intense storms.

Canyons

Canyons are narrow river valleys with near-vertical sides. They are caused by powerful fluvial erosion in deserts. Vertical erosion dominates as solid rock walls allow little sideways movement of river channels.

Pedestal rocks

Pedestal rocks are isolated, mushroom-shaped rocks. Zeugen is a collective term for rock pillars, rock pedestals and **yardangs** which have been undercut where less resistant rock underlies a layer of resistant rock, where sand grains are used in undercutting erosion and where moisture at the base of rocks is used in weathering processes.

Yardangs are streamlined parallel ridges of rock, aligned in the prevailing wind direction, caused by windblown erosion processes — abrasion.

Ventifacts

Ventifacts are exposed rocks lying on the desert surface that have been shaped by the erosion (abrasion) of windblown sediment in a sand-blasting effect.

Desert pavements

As fine material is removed, coarse material and pebbles are left behind, forming what is known as a desert pavement.

Typical mistake

Do not assume that all deserts are totally covered in sand dunes. Large areas of hot deserts are covered in rocky, stony surfaces.

The formation of distinctive depositional landforms

Dunes

Dunes are mounds and ridges of blown sand. To form they require an adequate supply of sand, strong and frequent wind and an obstacle to trap the sediment. Creep and saltation transport the sand up the windward slope, sand accumulates on the peak and eventually a small avalanche will occur down the slip face to restore equilibrium. In this way dunes advance in the direction of the prevailing wind. Figure 1.33 shows the formation of a dune. Where there is a large supply of sand and an expanse of dunes forms, it is referred to as an **erg** landscape.

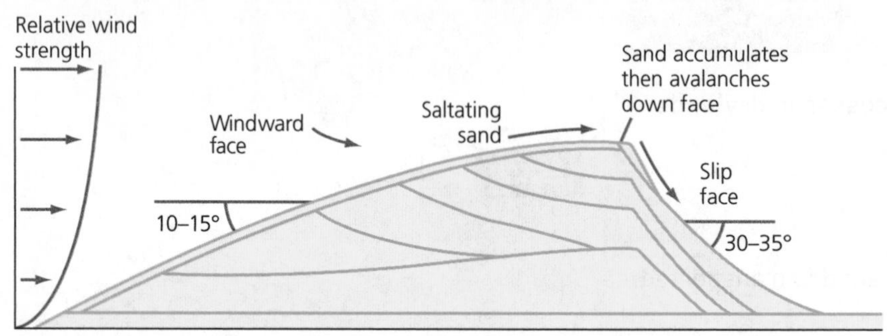

Figure 1.33 Formation of desert dunes

There are four main types of dune (Figure 1.34):

1 **Crescent dunes:** wider than they are long with a concave slip face. There are two subtypes: **barchans** (with horns that face downwind) and **transverse dunes** (a feature of large erg environments; they can be several hundred metres high).
2 **Linear or seif dunes:** straight or slightly curved. Often more than 100 km long and sometimes over 200 m high, with a slip face on alternate sides, they cover large areas in parallel, knife-edged ridges.
3 **Star dunes:** pyramidal in profile with slip faces on three or more sides. They form where the wind comes from different directions.
4 **Parabolic dunes:** have a 'U'-shaped form with arms that extend upwind.

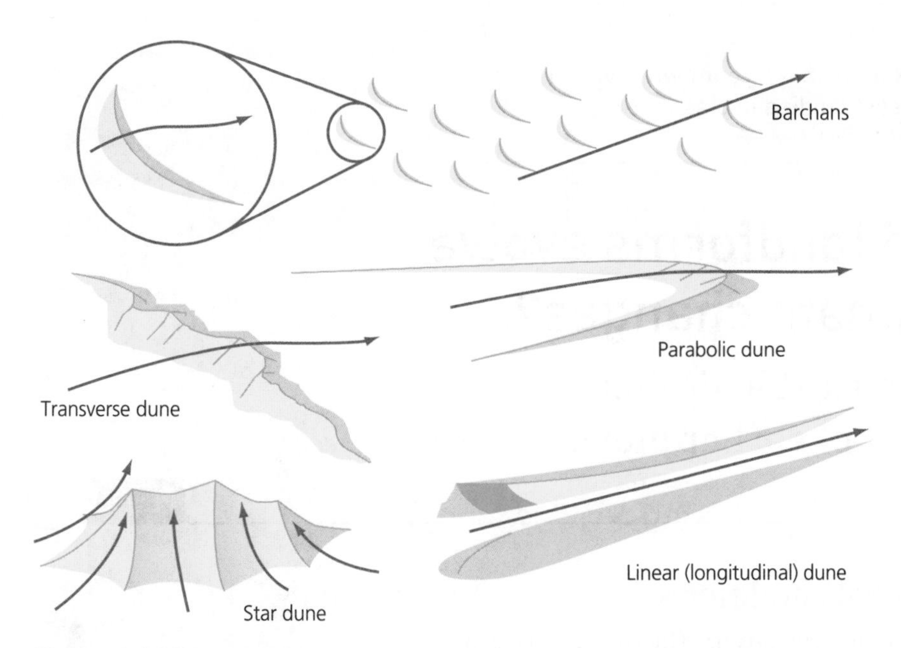

Figure 1.34 Dune types

Bajadas

Alluvial fans form where rivers leave steep-sided valleys (canyons) and enter adjacent lowlands. The reduction in gradient causes a sudden loss of energy and deposition. Where many alluvial fans develop, it is referred to as a bajada (pronounced: bahada).

Revision activity

Practise a clear sequence of explanation for **three** erosional and **three** depositional landforms.

Now test yourself TESTED

44 Outline what is meant by wind erosion and wind transport.
45 Name the **four** types of sand dunes.
46 How are zeugen formed?

Answers on p. 218

Exam tip

A quick sketch or an annotated diagram can be very effective in explaining physical landforms. Be confident in their use and make them **clear**.

Exam tip

In the past many desert environments had a more humid climate than today and so water had a past influence on desert landscapes which can still be seen today.

Exam tip

Landforms are small-scale features, while landscapes refer to the way features interconnect to form a landscape.

Dryland landforms are inter-related and together make up characteristic landscapes

REVISED

Case study of **one** mid-latitude desert and **one** low-latitude desert

Online (see p. 3) you will find a case study on Namib, southern Africa, to illustrate a low-latitude desert.

Revision activity

Based on the example of the Namib, make your own revision summary of a mid-latitude desert studied in class.

How do dryland landforms evolve over time as climate changes?

Fluvial landforms can exist in dryland landscapes as a result of earlier pluvial periods

REVISED

Climate change and pluvial conditions

Over the last 500,000 years climate in drylands has fluctuated between wet and dry conditions. Many present-day features are 'relict' features reflecting past climatic conditions. Wetter periods in the past are known as 'pluvials'.

Pluvial conditions, geomorphic processes and landforms

Weathering (particularly hydration and chemical weathering) and mass movement processes were more active in the pluvials. Rates of recession of **inselbergs** would have been more rapid and rivers would have transported large sediment loads.

An **inselberg** is an isolated, steep-sided mountain, often surrounded by extensive plains.

Modifications of landforms due to changes in climate

Because of more aridity in the past 6,000 years, weathering, mass movement and fluvial erosion in deserts have slowed. A decrease in vegetation and soil cover means that run-off has increased. Drier conditions have increased the effectiveness of aeolian erosion and transport. Future projections of climate change could mean that more aeolian erosion and transport will lead to deflation and advance of dunes. If extreme weather events increase, fluvial erosion could increase as a result of downpours.

Periglacial landforms can exist as a result of earlier colder periods

REVISED

Climate change in previous times and resultant colder conditions

Global climate in the past has fluctuated between cold and warm periods. The Pleistocene glacial (the most recent) reached its peak 20,000 years ago, approximately. Large parts of North America and Eurasia were covered in ice. Mid-latitudes were largely ice-free. Extreme cold, with temperatures averaging 6°C lower than at present, meant that **permafrost** areas dominated the landscape. Today some of these areas are deserts, e.g. the Great Plains.

Permafrost is permanently frozen soil and regolith.

Frost heave is the progressive movement of soil particles downslope as a result of freeze–thaw cycles.

Gelifluction or **solifluction** is the slow flow of saturated regolith down a gradient which can be very gentle.

Periglacial processes and landforms

The periglacial environment is free of ice sheets and glaciers but, apart from a shallow surface layer, the ground is permanently frozen. Important processes include **frost heave** and **gelifluction/solifluction**.

Characteristic landforms of these landscapes include:

- talus
- rock glaciers
- ognips or pingos
- cryoturbation (micro-scale: soil and sediment churned by frost heave)
- ice wedges
- nivation hollows

Figure 1.35 shows the combined features of periglacial landscapes.

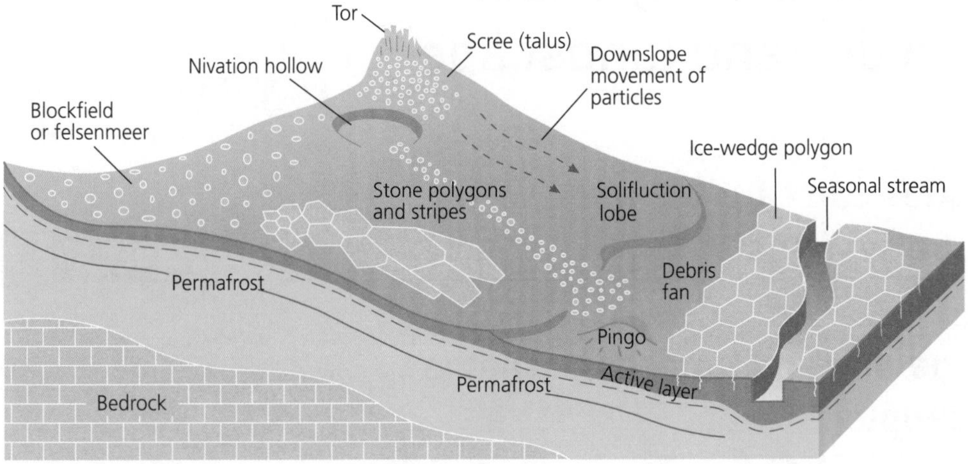

Figure 1.35 Features of periglacial landscapes

Exam tip

Read the specific command terms of exam questions carefully. Rarely will you be asked to just 'describe' a landform.

Revision activity

Make a list of the periglacial features listed in this section and complete a short summary of how each is formed. For example:

Talus: scree slopes developed by frost shattering

Now test yourself TESTED

47 Explain the process of frost heave.
48 Name the structural features of rock that make freeze–thaw weathering effective.

Answers on p. 218

Modifications of periglacial landforms due to changes in climate

Periglacial landforms in dryland landscapes form **relict** features. The processes that formed them are no longer operating. These landscape features will, over time, be modified, changed and eventually destroyed by present-day weathering, mass movement, fluvial and aeolian processes. An example is the effect of salt weathering and exfoliation in low latitudes which slowly disintegrates large boulders left in blockfields. Also, features such as sheets, lobes and terraces dry out and are eroded by **ephemeral** streams.

A **relict** is a remnant of a pre-existing formation.

An **ephemeral** stream is one that flows only briefly during or following a period of intense rainfall.

How does human activity cause change within dryland landscape systems?

Water supply issues can cause change within dryland landscape systems

REVISED

Case study of one dryland landscape that is being used by people

Revision activity

Select a suitable method of presentation, e.g. star diagram or annotated map, and summarise the case study covered in class for a dryland landscape that is being used by people. Focus on the following:

- the water supply issues taking place and the reasons behind them
- impacts on processes and flows
- the effect of these impacts in changing dryland landforms
- the consequence of these changes on the landscape

Economic activity can cause change within dryland landscape systems

REVISED

Case study of one dryland landscape that is being used by people

Revision activity

Select a suitable method of presentation, e.g. star diagram or annotated map, and summarise the case study covered in class for a dryland landscape that is being used by people. Focus on the following:

- the economic activity taking place and the reasons for it
- the impacts on processes and flows
- the effect of these impacts in changing dryland landforms
- the consequence of these changes on the landscape

Summary

- The systems approach can be applied to the study of dryland environments. Positive and negative feedbacks can lead to environmental issues, such as desertification in semi-arid environments which is the result of positive feedback.
- It is the interaction of a range of physical factors that forms dryland landscapes.
- Different types of dryland exist with a range of characteristics.
- Climatic and geomorphic processes interact to form characteristic dryland landforms and landscapes.
- Landscapes must be understood within specific locations/case studies.
- Changes to climate lead to dryland landscapes evolving and being changed over time, for example through the influence of pluvial conditions or colder climatic conditions.
- Human activity causes change within dryland landscape systems because of water supply issues and economic activity.
- Both water supply and economic activity issues must be understood in the context of a specific case study, with place-specific facts and reference to impacts and changing landscapes.

Exam practice

11 Using evidence from Figure 1.26, p. 45, suggest how the distribution of drylands is affected by climate. [3]

12 Suggest why latitude is important in the formation of tropical and sub-tropical drylands. [4]

13 Explain the contribution of moisture to weathering processes in dryland landscapes. [8]

14 Explain the formation of canyons. [8]

15 Using the data in Table 1.17:

(a) Calculate the interquartile range (IQR). [2]

(b) A fieldwork study of the relationship between time and average annual precipitation at Barrow gave a Spearman rank correlation result of R_s 0.766 for 20 paired values. Using Table 1.18, state the null hypothesis and interpret this result. [4]

Table 1.17 Average annual precipitation (mm) at Barrow, Alaska, 1995–2005

Average annual precipitation	81	74	131	99	121	146	132	89	131	183	139

Table 1.18

Degrees of freedom	0.05 significance level	0.01 significance level
19	±0.388	±0.549
20	0.377	0.534
21	0.368	0.521

Answers and quick quiz 1C online

ONLINE

2 Earth's life support systems

How important are water and carbon to life on Earth?

Water and carbon support life on Earth and move between the land, oceans and atmosphere

REVISED

The importance of water to life on the planet

- Oceans moderate temperatures by absorbing, storing and slowly releasing heat.
- Clouds (made up of water droplets) reflect approximately 20% of incoming solar radiation and lower surface temperatures.
- Water vapour (a greenhouse gas) absorbs long-wave radiation from the Earth, maintaining global temperatures almost 15°C higher than they would otherwise be.
- Water makes up 65–95% of all living organisms and is crucial to their growth, reproduction and metabolic functioning.
- Plants need water for photosynthesis, respiration and transpiration, and also to transport minerals from the soil and to maintain a rigid structure.
- Water is used for all chemical reactions in the body for humans and animals.
- Water is an essential economic resource for agriculture, manufacture and domestic purposes.

Carbon as the building block of life on Earth

Carbon is a chemical element. Stores include rocks, the atmosphere, oceans, sea floor sediments and the biosphere.

- Life is carbon-based — built on large molecules of carbon atoms such as proteins, carbohydrates and nucleic acids.
- Carbon is an economic resource as it is contained in fossil fuels.
- Agricultural crops and forest trees, which are used by humans in a variety of ways, also store carbon.

Typical mistake

When referring to carbon stores do not just focus on the atmosphere.

Water and carbon cycling

At a global scale water and carbon flow in closed systems (only energy and not matter crosses the boundaries) over time periods from days to millions of years. There is an important link between the flow of water and carbon and the atmosphere, oceans, land and biosphere. Figure 2.1 illustrates the relationship between the carbon and water cycle and the atmosphere.

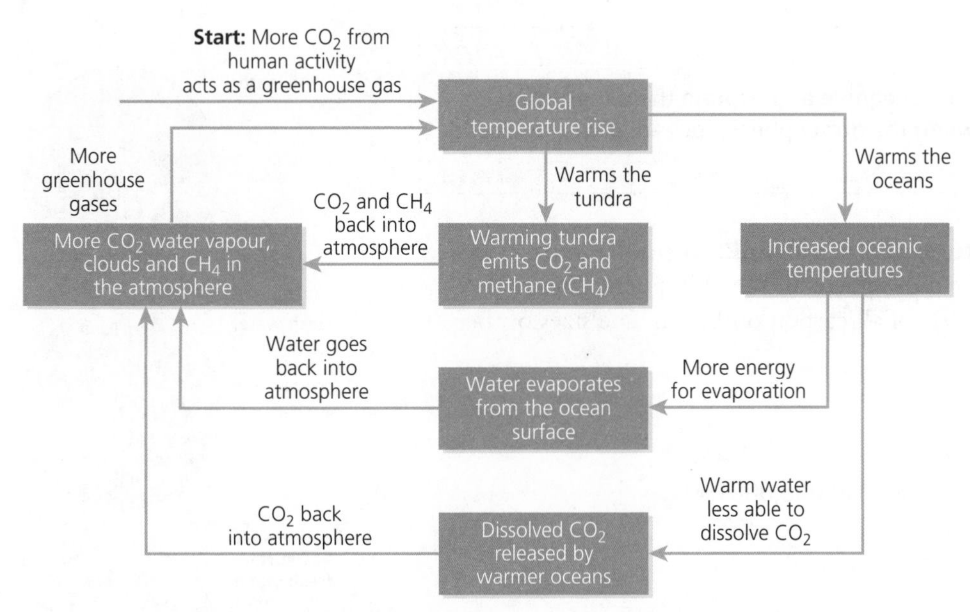

Figure 2.1 The link between the carbon cycle, the water cycle and the atmosphere

The carbon and water cycles are systems with inputs, outputs and stores

REVISED

The stores in carbon and water systems

The water cycle has three main types of water store: the atmosphere (where water exists as water vapour), oceans and land (cryospheric water and terrestrial water); these are summarised in Table 2.1. Water is held in these stores for varying amounts of time, e.g. a water molecule is held in the atmosphere for just nine days. About 71% of the Earth's surface is covered in water. The sizes of the world's land and ocean water stores are shown in Figure 2.2; note that the oceans contain 97% of global water. Water moves between these stores through a range of processes (see also Table 2.2, pp. 65–66):

- precipitation
- evapotranspiration
- run-off
- groundwater flow

Table 2.1 Types of water stores and global distribution

Type of water store	Description
Oceanic water	• There are five oceanic bodies of water and several smaller seas covering approximately 72% of the Earth's surface • The Pacific Ocean is the largest
Cryospheric water	• Composed of sea ice, ice caps, ice sheets, Alpine glaciers and permafrost • Mainly in high-altitude and high-latitude areas, including: ice sheets of Antarctica, Greenland, Arctic areas of Canada and Alaska; ice caps such as the Himalayas, the Rockies, the Andes and the southern Alps of New Zealand
Terrestrial water	• Rivers, the largest by discharge of water being the Amazon. Lakes — Canada and Finland have the largest number of lakes • Wetlands, where water covers the soil — these are present on every continent except Antarctica • Groundwater, soil water and biological water also make up terrestrial water
Atmospheric water	• The most common form is water vapour. Important as it absorbs and reflects incoming solar radiation. Warm air holds more water vapour than cold air — a small increase in water vapour will lead to an increase in atmospheric temperatures (positive feedback: see p. 71)

Exam tip

It is important to be able to recognise and explain the dynamic and cyclical relationship between the atmospheric, ocean and land-based water stores.

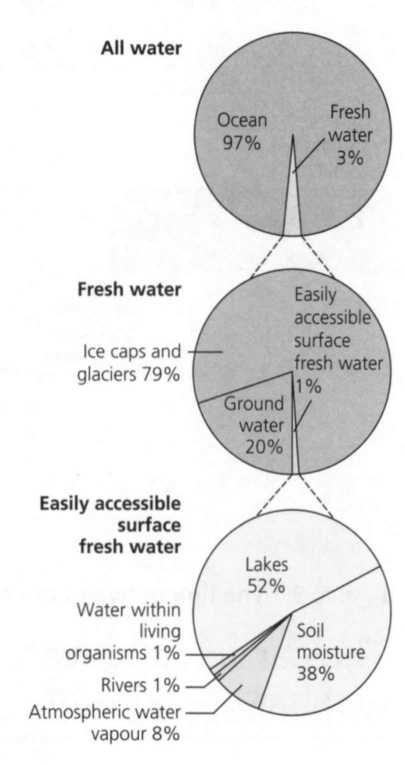

Figure 2.2 Sizes of land and ocean water stores

The carbon cycle has stores (known as **sinks** or **pools**) in rocks, sea floor sediments, oceans, the atmosphere and the biosphere (Figure 2.3). Sedimentary rocks hold 99.9% of all carbon on Earth. The sizes of the various stores are:

- atmosphere: 600 Gt
- ocean surface: 700 Gt
- ocean deep layer: 38,000 Gt
- sedimentary rocks: 60,000,000–100,000,000 Gt
- soil: 2,300 Gt
- terrestrial biomass: 560 Gt
- fossil fuels: 4,130 Gt

(Gt = Gigatonnes. 1 Gt is 1 billion tonnes)

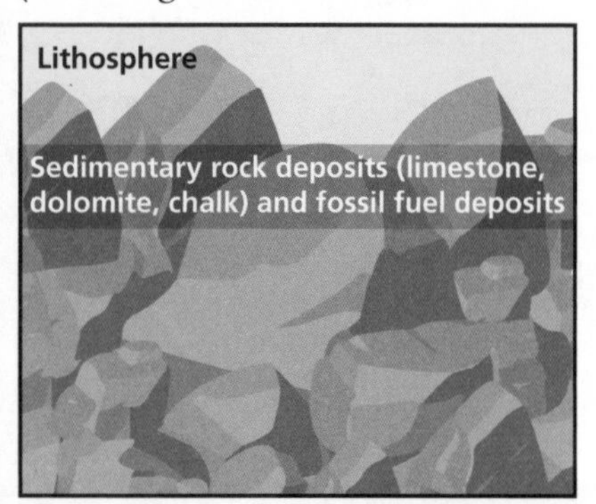

Hydrosphere

Oceans:

- Surface layer – photosynthesis by plankton
- Intermediate and deep layer – carbon passes through the marine food chains and sinks to the ocean bed, where it is decomposed into sediments
- Living and dissolved organic matter
- Calcium carbonate shells in marine organisms

Atmosphere

CO_2 gases in the atmosphere – a 'trace' gas accounting for 0.04% of the atmosphere, but this does not reflect its importance to life on earth and the fact that CO_2 is a potent greenhouse gas that plays a vital role in regulating the Earth's surface temperature

Figure 2.3 Major stores of carbon

The main processes leading to flows (or **fluxes**) between the carbon stores are as follows (see also Table 2.3, p. 67):

- photosynthesis
- respiration
- oxidation (decomposition and combustion)
- weathering

The characteristics of the water cycle

- Inputs and outputs of water form the **water cycle budget**. **Evaporation** from the oceans, soils, lakes and rivers and **transpiration** from plants make up inputs of water to the atmosphere. Together these processes are known as **evapotranspiration**.
- Moisture leaves the atmosphere as precipitation (rain, snow etc.) and **condensation** (fog). Water is released from ice cover by **ablation** and **sublimation**.

Exam tip

Note that on a global scale the water and carbon cycles are closed systems, but at a smaller scale, e.g. the drainage basin system, the system is open as both the sun's energy and materials cross system boundaries.

Flux is the rate of energy transfer per unit area.

The **water cycle budget** is the annual volume of water movement by, for example, precipitation, evapotranspiration and run-off, between stores such as oceans, permeable rock, vegetation and ice sheets.

Ablation is the loss of snow and ice through melting, evaporation and sublimation.

Sublimation is the change of water state from ice to vapour.

- **Run-off** transfers water from the land surface into rivers which flow into the sea. Some precipitation **infiltrates** the soil and becomes groundwater flow.
- Some water may **percolate** deeper into rock stores which are known as aquifers.

Run-off is the movement of water across the land surface.

Infiltration is the vertical movement of rainwater through the soil.

Percolation is the movement of surface and soil water into underlying permeable rock.

Water balance

The long-term balance between the inputs and outputs in a drainage basin system is known as the **water balance**. It is expressed in an equation as:

$$P = E + Q \pm S$$

where P is precipitation; E is evapotranspiration; Q is run-off (measured in river discharge); and S is change in storage.

- Positive water balance: precipitation exceeds evapotranspiration.
- Negative water balance: evapotranspiration exceeds precipitation.

Storage affects water balance. For example, in winter when precipitation is likely to be high, the soil storage may lead to a surplus of moisture and increased run-off. In summer utilisation of water by humans and vegetation is likely to be high and there may be a soil moisture deficit. In autumn, initial precipitation will recharge the soil store.

The characteristics of the carbon cycle

The carbon cycle describes the transfer of carbon from one store/pool to another. At its simplest level it is shown in Figure 2.4.

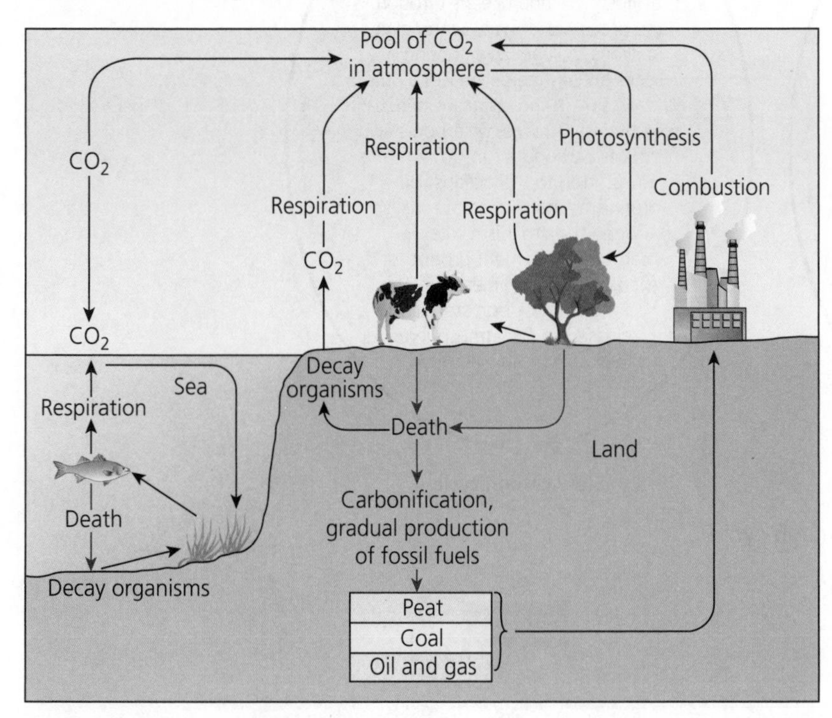

Figure 2.4 The carbon cycle

The carbon cycle has four sub-systems (Figure 2.5).

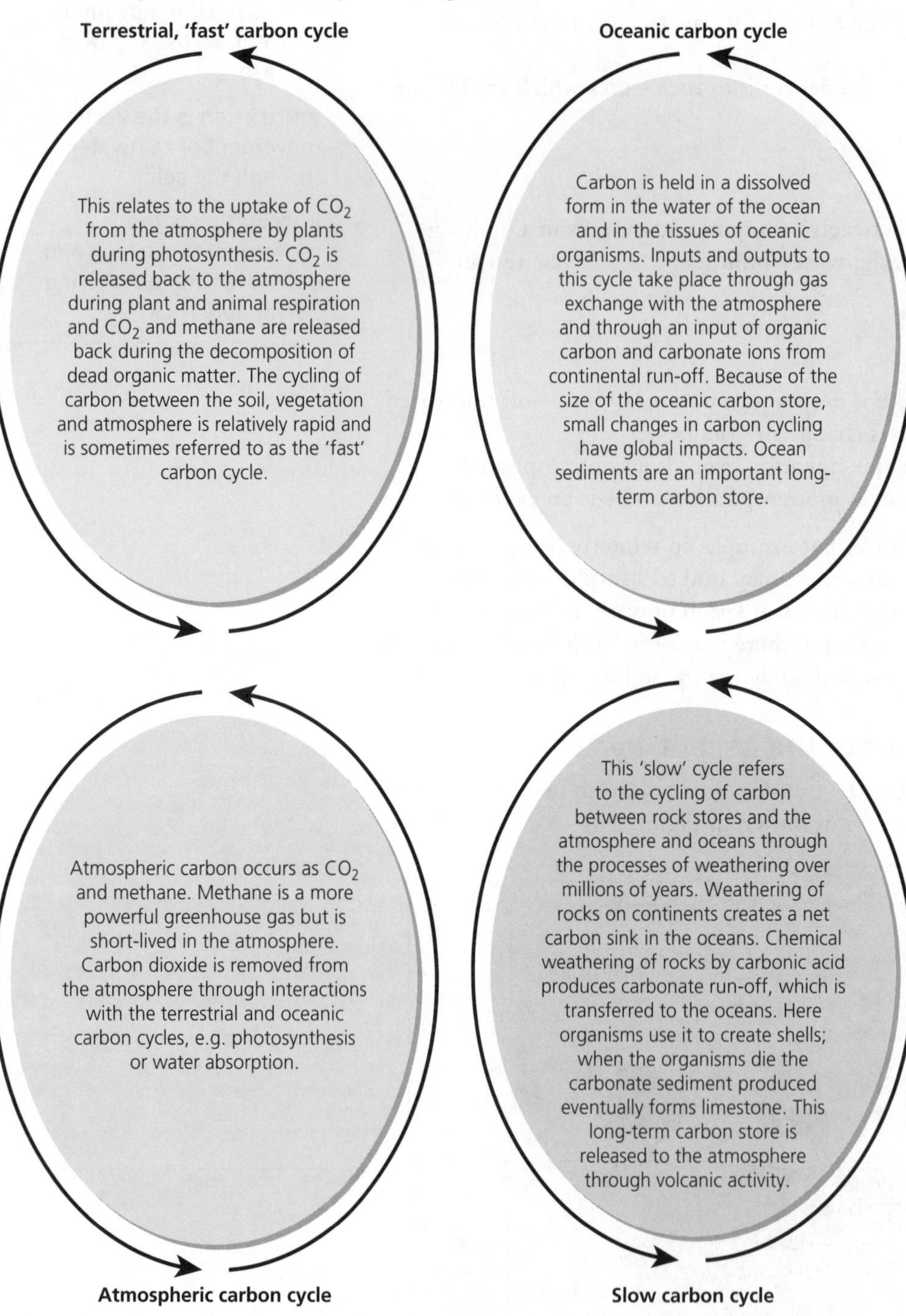

Figure 2.5 The sub-systems of the carbon cycle

Now test yourself

TESTED

1. Why do plants need water?
2. What are the main stores of carbon?
3. What processes move water between the different stores?
4. What processes move carbon between the different stores?
5. Explain the term water balance.

Answers on p. 218

The carbon and water cycles have distinctive processes and pathways operating within them

REVISED

Processes of the water cycle

Typical mistake

Do not ignore the concept of scale. All processes operate at scales ranging from individual slopes, to drainage basins to the global scale.

Water changes from one state to another, e.g. ice melts to form water (latent heat is needed), water freezes to form ice (latent heat is released). The following processes are key to an explanation of how water changes from one state to another:

- evaporation
- transpiration
- condensation
- cloud formation
- precipitation
- cryospheric processes

Evaporation and transpiration

Evaporation and transpiration are physical processes where liquid becomes gas. Heat energy is required, either provided by the movement of water or by solar energy and air which is not saturated and can therefore absorb evaporated water molecules (water vapour). Transpiration is linked to evaporation; it is a biological process where water is lost from plants through pores called stomata. Together the two processes are termed **evapotranspiration**. Factors affecting these processes include:

- temperature
- wind
- humidity
- climatic factors such as hours of sunshine

Revision activity

For each of the factors temperature, wind, humidity and climatic factors, write a sentence stating *how* they will affect both evaporation and transpiration.

Condensation

Condensation is a physical process where gas (water vapour) becomes liquid. It happens when air cools and is less able to hold water vapour (dew point). In the cooling process the water molecules condense onto nuclei (dust, smoke) or onto surfaces (e.g. grass), which forms water droplets or frost. Precipitation (rain, sleet, snow, hail) occurs when the air can no longer hold the condensed water.

Cloud formation

Clouds are visible masses of water droplets or ice crystals held in the atmosphere. They form when:

- air is saturated either because it has cooled below the dew point or evaporation means the air has reached its maximum water-holding capacity
- condensation nuclei are present

The greater the amount of moisture in the cooling air, the greater the condensation and cloud formation.

Causes of precipitation

Condensation that is a direct cause of precipitation can occur when:

1. air temperature is reduced to dew point, e.g. warm moist air passes over a cold surface on a clear night or heat is radiated out into the atmosphere and the ground gets colder, cooling the air above it
2. the volume of air increases as it rises and expands but there is no addition of heat (adiabatic cooling). In this example the air may be forced to rise for three different reasons, each resulting in precipitation:
   - (i) air is forced to rise over hills and mountains, producing **orographic rainfall**
   - (ii) air masses of different temperatures and densities meet; the warm air rises over the cool sinking air and results in **frontal rainfall**
   - (iii) warm air rises from hot surfaces on a sunny day causing **convectional rainfall**

Cryospheric processes

Cryospheric processes affect the mass of ice at any scale, and include:

- **accumulation:** inputs to a glacial system from snowfall
- **ablation:** output from a glacial system due to melting

At a global scale these processes often occur in cycles of glacial periods and interglacial periods. A third process is:

- **sublimation:** ice changing directly into water vapour

Revision activity

Construct a diagram to summarise the processes involved when water changes state.

Now test yourself

TESTED

6. What is the difference between an open and a closed system?
7. Explain how the processes of evaporation and condensation relate to the formation of:
   (i) clouds
   (ii) rainfall.
8. How do the following affect evapotranspiration: temperature, wind, humidity?
9. Why do sunny days lead to bursts of heavy rainfall?

Answers on pp. 218–9

Catchment hydrology

A **drainage basin** is an area of land drained by a river and its tributaries. The drainage basin water cycle:

- forms a sub-system of the hydrological or water cycle
- is an open system as it has inputs and outputs of both matter and energy
- is composed of **inputs** (precipitation), **flows and transfers** (throughfall, stemflow, infiltration, percolation, overland flow and groundwater flow) and **outputs** to the sea or atmosphere (evapotranspiration) (Figure 2.6; see Table 2.2 (pp. 65–66) for key terms).

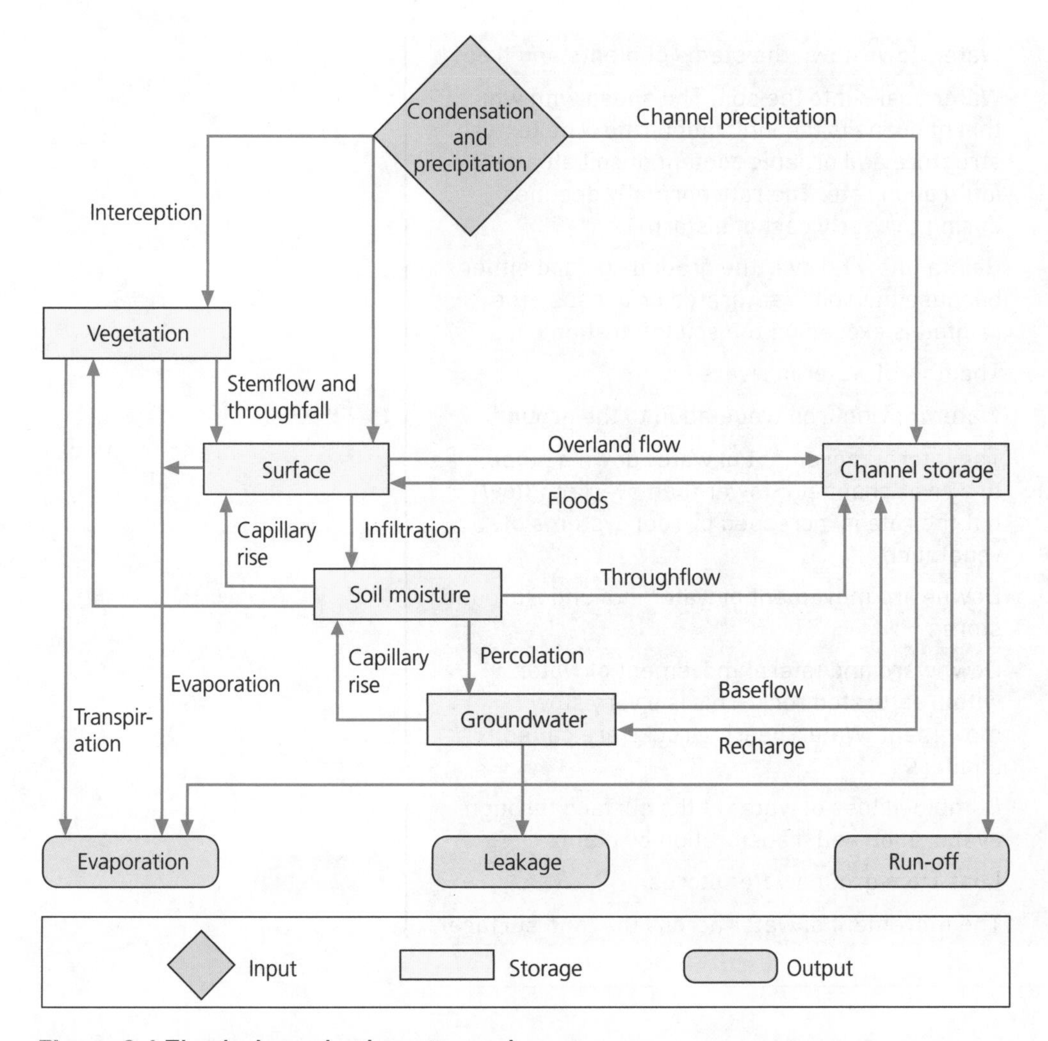

Figure 2.6 The drainage basin water cycle

Table 2.2 Key terms for the drainage basin water cycle

Precipitation	May fall as rain, hail, sleet or snow. The duration and intensity will impact processes within the system	**Inputs**
Condensation	Water vapour turning to liquid. May form fog	**Inputs**
Vegetation store (or interception store)	Vegetation cover intercepts the precipitation and a store may be held on leaves and branches. Density of vegetation will affect this. Tropical rainforest can intercept 58% of rainfall	**Stores**
Surface storage	This mainly occurs in built environments as puddles. In natural environments infiltration normally occurs faster than rainfall and there will only be surface puddles after very long periods of rainfall or on impacted surfaces or bare rock	**Stores**
Soil moisture/soil water storage	Pore spaces between soil particles fill with air and water. The amount of pore space varies in different soils: clay has 40–60% volume; sand has 20–45%	**Stores**
Groundwater store	Water stored underground in permeable and porous rocks	**Stores**
Channel store	The volume of water in a river channel	**Stores**

Stemflow	Water flows down the stems of plants and trees	Flows
Infiltration	Water soaks into the soil. The speed with which this happens is the infiltration rate. The texture, structure and organic content of soil all affect infiltration rate. The rate normally declines during the early part of a storm	
Overland flow	Rainfall flowing over the ground surface either because the soil is saturated or because the rainfall is exceeding the soil infiltration capacity	
Channel flow	The flow of water in rivers	
Throughfall	Water moving from vegetation to the ground	
Throughflow	The lateral movement of water down a slope to a river channel. Slower than overland flow but the rate is increased by root systems of vegetation	
Percolation	Downward movement of water into underground stores	
Groundwater flow	Downward and lateral movement of water within saturated rock. This is a very slow movement. Water-bearing rocks are called aquifers	
Evapotranspiration	Combined loss of water at the surface through evaporation and transpiration by plants	Outputs
Leakage	Loss from groundwater stores	
Run-off	The movement of water across the land surface	

Revision activity

In pairs, using Table 2.2, test each other on the key terms: either give the term and ask for an explanation or give the explanation and ask for the term to which it refers.

Typical mistake

Drainage basin water cycles are not static — they are in a constant dynamic state because of climatic change.

Exam tip

The use of the correct terms in extended writing explanations will gain credit.

Processes of the carbon cycle

The main processes of the carbon cycle include photosynthesis, respiration, decomposition, combustion (including natural and fossil fuel use), sequestration in the oceans and weathering. They are summarised in Table 2.3.

Table 2.3 **Processes of the carbon cycle**

Process	Description
Precipitation	Atmospheric CO_2 dissolves in rainwater to form weak carbonic acid. Rising concentrations of CO_2 in the atmosphere resulting from **anthropogenic** emissions have increased the acidity of rainfall and led to increased acidity of ocean surface waters
Photosynthesis	Plants use energy from sunlight and combine CO_2 from the atmosphere with water from the soil to form carbohydrates. Virtually all organic matter is formed from this process. Carbon is stored (or sequestered) for long periods of time because trees can live for hundreds or thousands of years and resistant structures such as wood take a long time to decompose
Respiration	Plants release CO_2 back to the atmosphere through respiration (about half of the terrestrial portion) Soil respiration: microscopic organisms living in soil also release CO_2 through respiration
Decomposition	The process of decomposition by fungi and bacteria returns CO_2 to the atmosphere. Decomposition also produces soluble organic compounds dissolved in run-off from the land surface. Greenhouse gases (GHG) are released as a by-product
Combustion	Fossil fuels (coal, oil and natural gas) contain carbon captured by living organisms over periods of millions of years and stored in the Earth's crust. Since the Industrial Revolution these fuels have been mined and combusted to serve as a primary energy source. The main by-product of fossil fuel combustion is CO_2
Carbon **sequestration** in oceans and sediments: the oceanic carbon pump and the biological pump	CO_2 moves from the atmosphere to the ocean by the process of diffusion. At low latitudes warm water absorbs CO_2. At high latitudes where cold water sinks, the carbon is transferred deep into the ocean. Where the cold water returns to the surface and warms again, it loses CO_2 to the atmosphere; in this way CO_2 is in constant exchange between the oceans and the atmosphere. This vertical circulation is a process called the **oceanic carbon pump**. Phytoplankton also fix CO_2 through photosynthesis, and the carbon passes through the oceanic food web. Carbonate is removed from the sea by shell-building organisms. When organisms die the shells sink into deep water; the decay of marine organisms releases some carbon dioxide into the deep water (the **biological pump**). Some material forms layers of carbon-rich sediments which over millions of years turn to sediments in rocks
Weathering	Weathering processes (driven by the atmosphere, rain and groundwater) break down rocks on the Earth's surface. These small particles are combined with plant and soil particles and eventually carried to the ocean. Large particles are deposited on the shore. The sediment accumulates. Layers build and eventually, because of surface pressure, shale rock is formed. Within the ocean, dissolved sediments mix with the seawater and are used by marine organisms to make skeletons and shells containing calcium carbonate. When these organisms die the carbonate collects at the bottom of the ocean and sedimentary rocks (e.g. limestone) form

Exam tip

You must be clear on the term **processes**. It is also important to understand the factors that **affect the rate** of those processes, e.g. increased sunlight will lead to increased photosynthesis (and consequently greater uptake of water and CO_2 by plants).

Anthropogenic: involving humankind as a driver of environmental change.

Sequestration is the long-term storage of carbon dioxide and other forms of carbon to mitigate climate change.

Now test yourself

TESTED

10 One pool of CO_2 is in the atmosphere. What processes transfer CO_2 to the atmosphere?
11 Combustion releases carbon into the atmosphere. What is the source of this carbon?
12 What is the oceanic carbon pump?
13 Draw a simple flow diagram of the oceanic carbon pump.
14 How can warm and cold climates affect decomposition rates on land?
15 What is carbon sequestration?

Answers on p. 219

How do the water and carbon cycles operate in contrasting locations?

The physical and human factors affecting the water and carbon cycles in the tropical rainforest

REVISED

Case study of a tropical rainforest

Online (see p. 3) you will find a case study on the Amazon rainforest to illustrate a tropical rainforest.

The physical and human factors affecting the water and carbon cycles in an Arctic tundra area

Case study of the Arctic tundra

Revision activity

Using the layout of the Amazon rainforest case study, make a revision summary of your case study of the Arctic tundra. Focus on the following headings:

- Water and carbon cycles specific to the Arctic tundra.
- Physical factors affecting flows and stores in the water cycle.
- Physical factors affecting flows and stores in the carbon cycle.
- Seasonal changes in the water and carbon cycles in the Arctic tundra.
- The impact of the oil and gas industry on the water and carbon cycles.
- Management strategies to moderate the impact of the oil and gas industry.

Exam tip

These two case studies form a large section of the unit on Earth's life support systems and provide a context to apply the theoretical content. Make sure that you can provide detailed information on the elements required, including place-specific facts.

How much change occurs over time in the water and carbon cycles?

Human factors can influence the processes and stores in the water and carbon cycles

REVISED

Dynamic equilibrium in the cycles

The inputs, throughputs, outputs and stores in the water and carbon cycles are in a constant changing state called **dynamic equilibrium**. There will be short-term fluctuations and in the long term a balance is maintained. The balance of the system is restored by **negative feedback loops**.

Revision activity

Give an example of negative feedback for both the water and carbon cycles and represent the explanation as a flow diagram.

Land use change and water extraction

Table 2.4 outlines different types of human impact on the water cycle (blue text) and carbon cycle (red text).

Table 2.4 Human impact on the water and carbon cycles

Urbanisation	Forestry
• Natural surfaces: vegetation and soil replaced by impermeable concrete, brick and tarmac surfaces • Infiltration reduced • Urban drainage systems remove surface water rapidly, e.g. gutters • Water levels in rivers rise rapidly owing to quick transfer of surface water • In particular, urban development on floodplains reduces water storage capacity and leads to increased river flow and flooding • Urban growth reduces the amount of surface vegetation • CO_2 emissions from energy consumption, transport and industry increase • Increase in CO_2 emissions from cement manufacture	• Plantations of natural forest increase interception of rainfall, e.g. conifers which are evergreen and planted at high density • Evaporation increases as leaf store water evaporates directly back to the atmosphere • Run-off and stream discharge is reduced • Lag times are long, peak flow low and total discharge low in plantation areas • Transpiration in forested areas is higher than for farmland and moorland • Localised deforestation means that evapotranspiration is lower as new minimal vegetation cover has fewer leaves and fewer roots; there is less interception because of reduced canopy, overland flow and throughflow increase as there is a lack of vegetation to slow down these processes. There is increased river discharge and risk of localised flooding • Changing land use to forestry increases carbon stores • Forest trees extract CO_2 from the atmosphere and sequester it for hundreds of years. Most of the carbon is stored in the wood of the tree stem • Forest trees are only an active **carbon sink** for the first 100 years or so after planting and so forestry plantations usually have a rotation of 80–100 years

➜

Farming practices	Water extraction
• Irrigation diverts water from rivers and groundwater supplies to cultivated land. Some of this water is used by plants from soil storage and released by transpiration • Interception, evaporation and transpiration are all lower in agroecosystems than in forest and grasslands ecosystems • Ploughing increases soil moisture loss and can form drainage channels which increase run-off and lead to soil erosion • Underground drainage channels in farmland increase water transfer to rivers • Use of heavy machinery can compact the soil and increase run-off • Clearance of forest for farming reduces above- and below-ground carbon stores • Ploughing reduces soil carbon storage and exposes soil organic matter to oxidation • Harvesting means that only small amounts of organic matter are returned to the soil, further reducing carbon stores • Rice paddies generate methane • Livestock release methane gas as a by-product of digestion • Emissions from tractors increase the level of CO_2 in the atmosphere	• Water extraction is the process of taking water from a surface or ground source either temporarily or permanently • Uses can be agricultural, industrial or domestic • Hydrogeology is used to monitor safe levels of water extraction as over-extraction can lead to several issues including: o rivers drying up o damage to wetland ecosystems o sinking water tables (if extraction exceeds recharge) o empty wells • In coastal areas intrusion of salt water from the sea degrades groundwater and leads to difficulties of usage for domestic and agricultural purposes • **Aquifers:** water-bearing rocks include chalk and sandstone. Groundwater is abstracted for use by wells and boreholes. The border between saturated and dry rock is the water table and this fluctuates according to season and amount of water abstraction • **Artesian basins:** sedimentary rocks may form a basin shape or 'syncline'; an aquifer trapped between impermeable rock layers may contain groundwater which is under artesian pressure. A well could allow the water to flow to the surface under its own pressure — an artesian aquifer

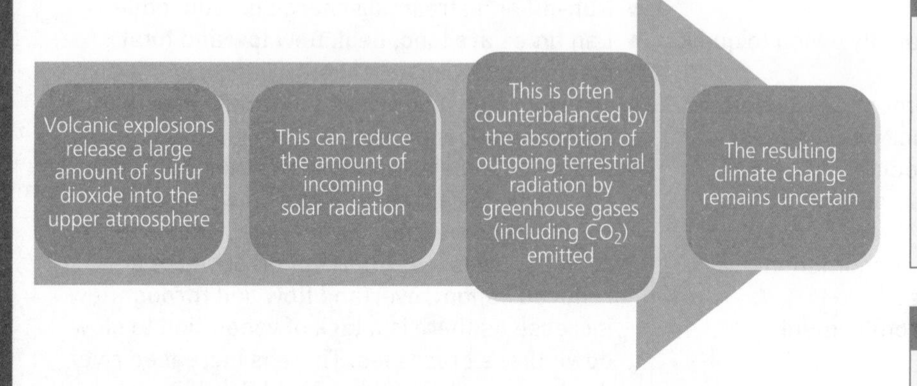

Figure 2.7 The impact of volcanic activity on the carbon cycle

Exam tip

Forest can be cleared for farming by controlled fires which release plant carbon into the atmosphere, but forest fires can also be naturally induced in periods of intense drought.

Exam tip

Make sure that you are able to refer to an example of water extraction from a river catchment and extraction from a named aquifer.

Revision activity

In small groups, divide the human impacts on the water and carbon cycles in Table 2.4 between you, e.g. three people each take two impacts. Produce a flow diagram that explains the cause and effect of each impact and explain this to the rest of the group.

An example is shown in Figure 2.7 for the potential impact of volcanic activity on the carbon cycle, but remember this is a *natural* not a human impact.

Fossil fuel combustion

There remains a high global dependency on fossil fuels (coal, oil and natural gas). Approximately 10 billion tonnes of CO_2 are released into the atmosphere annually. Anthropogenic carbon emissions impact significantly on the carbon stores of the atmosphere, oceans and biosphere. Combustion of fossil fuels and the resulting transfer of carbon from geological stores to the atmosphere and oceans is the main cause of global warming. One possible solution is carbon capture and storage (CCS). However, this is limited by high capital costs and by the fact that the process uses large amounts of energy and requires storage reservoirs with specific geological conditions.

Positive and negative feedback in the water and carbon cycles

Feedback is a natural response to a change in a system's equilibrium. The feedback can be **positive** (when the initial change causes further change) or **negative** (the system change is countered and equilibrium is restored). Sometimes one factor can induce both positive and negative feedback, e.g. atmospheric water vapour; see Table 2.5.

Exam tip

The idea of positive and negative feedback can be applied to a range of concepts. The same catalyst can lead to both positive and negative feedback, e.g. water vapour.

Revision activity

Complete Table 2.5 to give a cause-and-effect explanation sequence of the different examples of positive and negative feedback within and between the carbon and water cycles. The first one has been done for you.

Table 2.5 Positive and negative feedback in the water and carbon cycles

Example	Type of feedback	Explanation
The effect of rising temperatures on the water cycle	Positive feedback	Warmer temperatures → more evaporation and the atmosphere holds more water vapour → more cloud and precipitation → more evaporation and water vapour held in the atmosphere → more absorption of long-wave radiation → warmer temperatures
The effect of more atmospheric vapour on the water cycle	Negative feedback	
Above-average precipitation in a drainage basin water cycle	Negative feedback	
Excessive CO_2 in the atmosphere	Negative feedback	
Impact of global warming on the carbon cycle	Positive feedback	

Variations in the water and carbon cycle pathways and processes over time

REVISED

Given the debate on climate change and the potential damage it could cause, there is close monitoring through satellite technology and remote sensing of global air temperatures, sea surface temperatures, sea ice thickness and rates of deforestation. This continuous monitoring allows data to be analysed for indications of both short- and long-term change.

Short-term changes

Diurnal changes

Significant changes can occur within 24 hours in the water cycle. Evaporation and transpiration are both much lower at night as temperatures drop. Downpours in the afternoon because of intense convectional heating are a feature of some global climates.

Carbon flows from the atmosphere to vegetation during the day; the flux is reversed at night. Low levels of sunlight reduce photosynthesis in vegetation on land and in phytoplankton in oceans.

Seasonal changes

Seasons are controlled by variations in the intensity of solar radiation. This has an impact on rates of evapotranspiration and precipitation which impact the water cycle.

Seasonal variations in the carbon cycle are shown in month-to-month variations in **net primary productivity** (NPP). During the summer months in the Northern Hemisphere, for example, there is a net flow of CO_2 from the atmosphere to the biosphere as vegetation is in full foliage and photosynthesis is rapid.

Net primary productivity (NPP) is energy produced by plants, taking into account energy used for respiration.

Long-term changes

Over the last 1 million years the global climate has been unstable, with large fluctuations in temperatures occurring at regular intervals. In the past 400,000 years, there have been four major glacial cycles, with cold glacials followed by warmer interglacials. Figure 2.8 shows the impacts of long-term climate change.

Impact of glacials on the water cycle

Sea levels fall
Ice sheets and glaciers expand
Ice sheet advance destroys forest and grassland
Water stored in the biosphere shrinks
Evapotranspiration declines
Water cycle slows due to reduction in evapotranspiration and water storage

Impact of glacials on the carbon cycle

Less carbon dioxide in the atmosphere
Changes in oceanic circulation bring nutrients to the surface; phytoplankton grows rapidly and fixes carbon dioxide in photosynthesis; when phytoplankton die the carbon is stored in the deep ocean
Less exchange of carbon between the soil and the atmosphere due to ice coverage
Because of the increase in ice coverage there is less vegetation and a reduction in the carbon fixed by photosynthesis

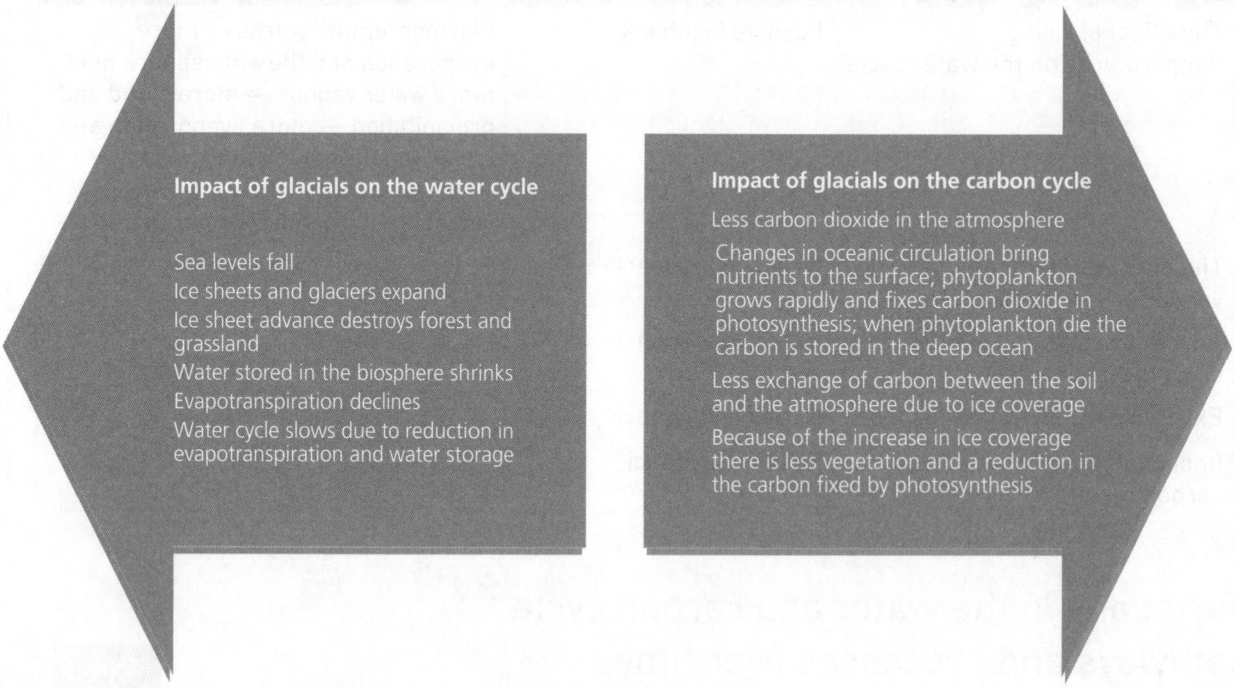

Figure 2.8 Impacts of long-term climate change on the water and carbon cycles

The importance of research and monitoring

- Cycling of carbon and water are central to supporting life on Earth and an understanding of these cycles and how they are changing is central to managing global challenges such as the impacts of climate change and consequences for future water, food and energy supply.

- Changes in the water and carbon cycles are central to analysis of environmental change and the global challenges presented.
- Understanding of the regional variations in the sources and sinks of CO_2 helps identify sequestration and emission management options.

To what extent are the water and carbon cycles linked?

The two cycles are linked and interdependent

REVISED

Links and interdependence in the two cycles

The water and carbon cycles interact directly where carbon is transported dissolved or suspended in running water. Transport of weathering products and organic matter from the continents to the oceans is an important aspect of carbon cycling which is directly linked to water flux. Similarly, the impact of changing atmospheric carbon concentrations on global climate has a profound effect on water cycling, impacting terrestrial and oceanic evaporation and patterns of precipitation (Figure 2.1, p. 59).

The two cycles are also linked through the role of ecosystems in carbon cycling since moisture availability is a key control on plant distribution and plant life plays a key role in terrestrial carbon cycling.

Climate change and land-use change may lead to significant change in the functioning of terrestrial ecosystems which impact on both water and carbon cycling; interlinks and inter-dependency are summarised in Table 2.6.

Table 2.6 How the water and carbon cycles are interlinked and interdependent

	Interlinkages
Atmosphere	Atmospheric CO_2 has a greenhouse effect. CO_2 plays a vital role in photosynthesis by terrestrial plants and phytoplankton. Plants, which are important carbon stores, extract water from the soil and transpire it as part of the water cycle. Water is evaporated from the oceans to the atmosphere, and CO_2 is exchanged between the two stores
Oceans	Ocean acidity increases when exchanges of CO_2 are not in balance (i.e. inputs to the oceans from the atmosphere exceed outputs). The solubility of CO_2 in the oceans increases with lower sea-surface temperatures (SSTs). Atmospheric CO_2 levels influence: SSTs and the thermal expansion of the oceans; air temperatures; the melting of ice sheets and glaciers; and sea level
Vegetation and soil	Water availability influences rates of photosynthesis, NPP, inputs of organic litter to soils and transpiration. The water-storage capacity of soils increases with organic content. Temperatures and rainfall affect decomposition rates and the release of CO_2 to the atmosphere
Cryosphere	CO_2 levels in the atmosphere determine the intensity of the greenhouse effect and melting of ice sheets, glaciers, sea ice and permafrost. Melting exposes land and sea surfaces which absorb more solar radiation and raise temperatures further. Permafrost melting exposes organic material to oxidation and decomposition which releases CO_2 and CH_4. Run-off, river flow and evaporation respond to temperature change

How human activities cause change in the availability of water and carbon

The human activities causing change in the availability of water and carbon are summarised in Figure 2.9.

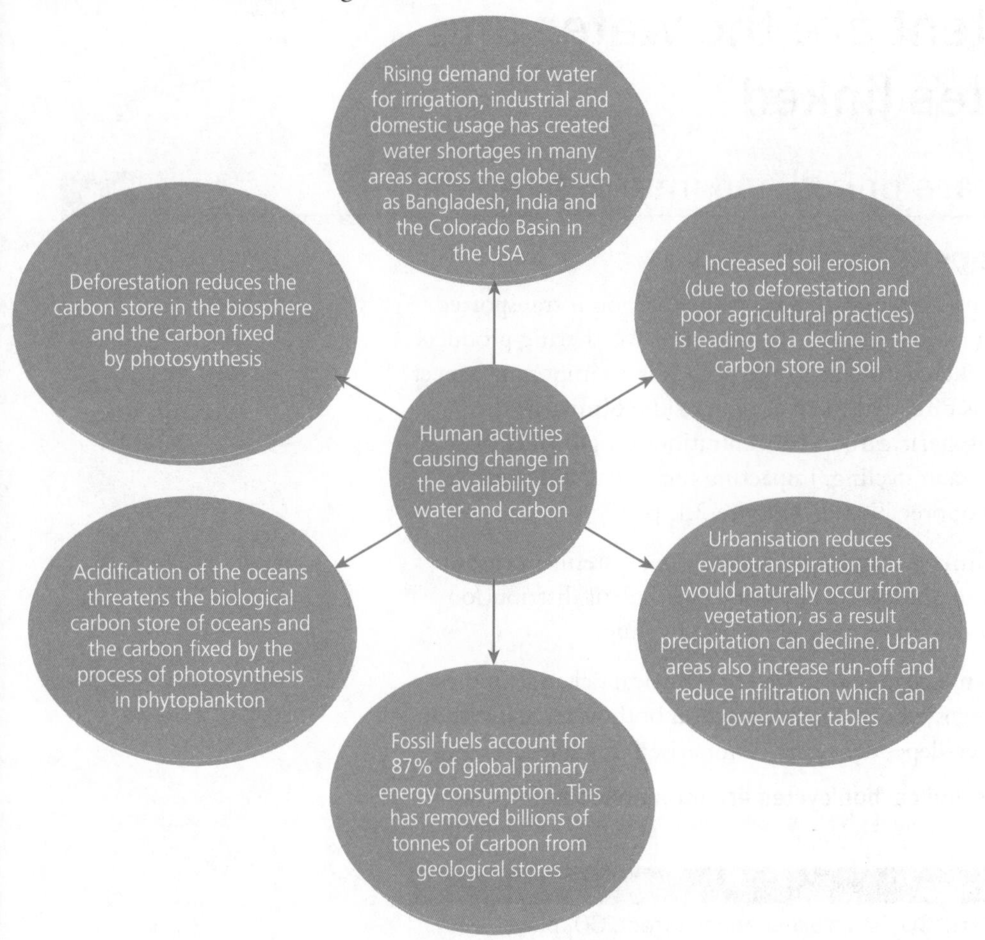

Figure 2.9 How human activities cause change in the availability of water and carbon

Figure 2.10 shows in percentage terms how human activities such as deforestation and urbanisation reduce evapotranspiration and precipitation, increase run-off, decrease throughflow and lower water tables.

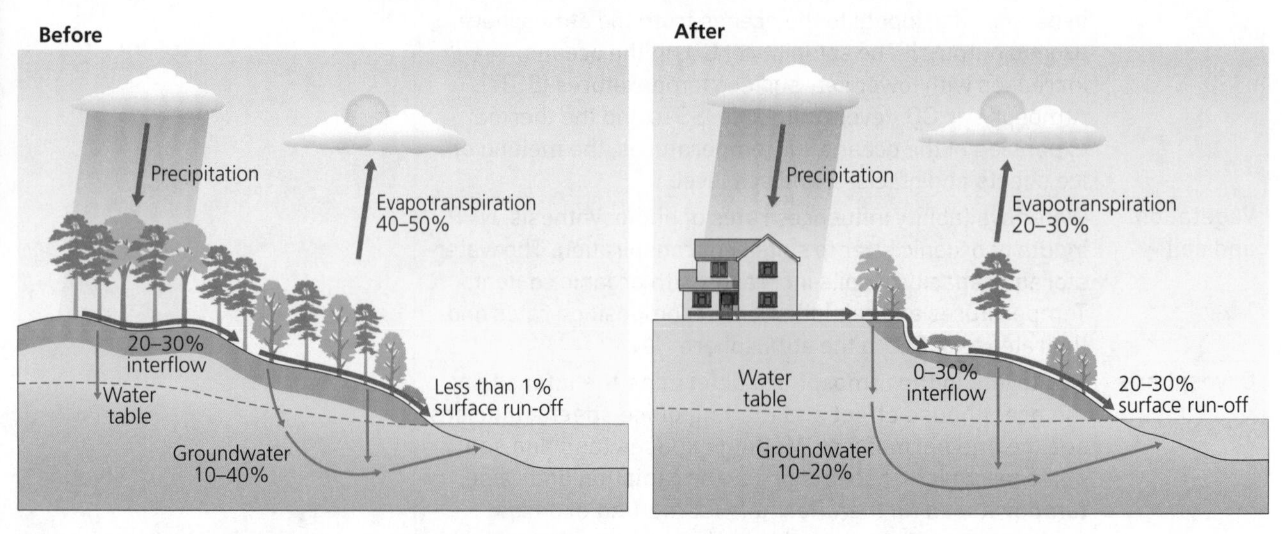

Figure 2.10 Changes in the water cycle caused by human activity

The impact of long-term climate change on the water and carbon cycles

Water cycle

- Global warming has increased the amount of evaporation and thereby the amount of water vapour in the atmosphere. The positive feedback of water vapour — a greenhouse gas — further increases global temperatures, evaporation and precipitation.
- Increased precipitation in areas where there is urbanisation, building on floodplains and deforestation will increase flood risks.
- Water vapour is a source of energy in the atmosphere which can lead to more extreme weather events, e.g. storms and tornadoes.
- Water stored in the cryosphere will shrink as global warming increases melting of glaciers and will be transferred into oceans.

Carbon cycle

- Impacts are more complex, but the long-term impact of climate change will probably be an increase in carbon stored in the atmosphere, a decrease in carbon stored in the biosphere and a decrease in the carbon store of oceans.
- Higher temperatures increase rates of decomposition, and the rate of carbon transfer from the biosphere and soil to the atmosphere will increase.
- Where temperatures are so high that aridity increases, forests will be replaced by grasslands which reduces the carbon store in woody vegetation.
- Global warming may allow boreal forests to spread north.
- In permafrost areas carbon is being released from frozen ground as temperatures rise.
- Ocean acidification (caused by CO_2 dissolving in oceans and creating carbonic acid) is limiting the capacity of oceans to store carbon.

Now test yourself TESTED

16 Why is water vapour considered such a dangerous greenhouse gas?
17 How is ocean acidification leading to reduced carbon stores in oceans?

Answers on p. 219

The global implications of water and carbon management

REVISED

Exam tip

Remember that impacts can always be categorised into human and physical, and the human into economic, political and socio-cultural.

Exam tip

Impacts are often complex in nature and it is difficult to predict the precise rate, magnitude and direction of change.

Global management strategies to protect the carbon cycle

Table 2.7 Management strategies to protect the carbon cycle

Afforestation
• This is the planting of trees in areas where deforestation has taken place or in new areas • Trees act as an important carbon sink and so they can reduce atmospheric CO_2, and reduce flood risks and soil erosion • The UN's Reducing Emissions from Deforestation and Forest Degradation (REDD) schemes encourage developing countries to conserve their forests
Wetland restoration
• Wetlands account for 35% of the terrestrial carbon pool and include marshes, peatlands, floodplains and mangroves • They form important carbon sinks and protection schemes include the **International Convention on Wetlands** and the **European Union Habitats Directive** • Restoration at a local level involves raising water tables to create waterlogged conditions. Water levels can be maintained also by diverting or blocking drainage ditches and installing sluice gates
Improved agricultural practices
• Mulching adds organic matter and prevents carbon losses from the system • Rotation of cash crops with cover crops can increase the biomass returned to the soil • Improved crop varieties can increase productivity and enhance soil organic carbon (SOC)
Reducing emissions
• **Carbon trading:** businesses can be allocated a quota for CO_2 emissions; **carbon credits** are received for emissions lower than the quota and financial penalties or the opportunity to purchase additional credits are the consequences of exceeding the quota. **Carbon offsets** are credits to countries and businesses for schemes such as afforestation and use of renewable energy • **International agreements:** as climate change is a global problem, international cooperation is key to reducing carbon emissions. However, reaching consensus and action from all countries is complex and frequently affected by the self-interests of different countries, industries and organisations. The only significant global agreement has been the **Kyoto Protocol (1997)** where most rich countries agreed legally binding reductions in CO_2 emissions. Even so, India and China were exempt and the USA did not ratify the agreement. After Kyoto expired in 2012 a new agreement was reached in the **Paris Climate Convention 2015** for implementation in 2020. Countries will set their own voluntary targets which are not legally binding • **Transport innovations:** attempts to reduce greenhouse gas emissions from road and aviation transport form a key element of mitigation. Road transport initiatives include sustainable transport schemes, congestion charging (London), park and ride schemes (Cambridge) and integrated transport networks (Curitiba, Brazil); there is a range of mitigation strategies within the aviation industry, e.g. adopting fuel-efficient routes and cruising at a lower speed

Global management strategies to protect the water cycle

Revision activity

Complete Figure 2.11 and make bullet-point notes on the strategies used to protect the water cycle (drainage basin planning and water allocations). Forestry has been done for you.

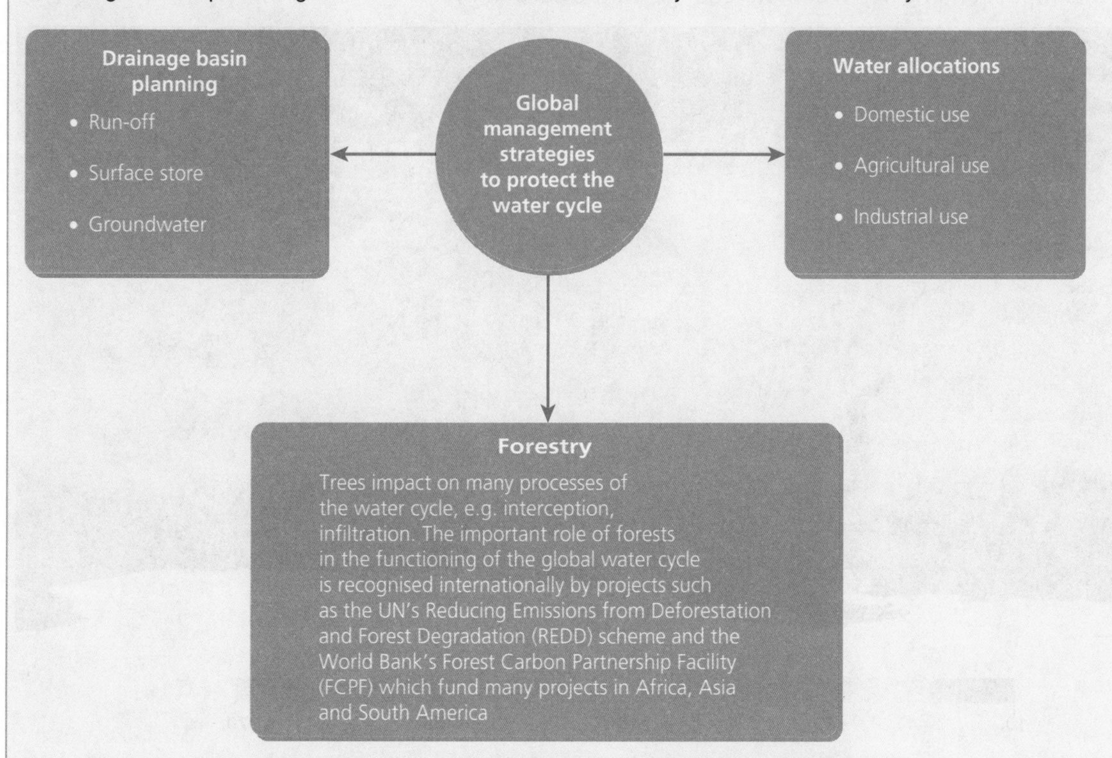

Figure 2.11 Revision notes on global management strategies to protect the water cycle

Exam tip

There is a wide range of strategies to protect the carbon and water cycles. It is important to be able to evaluate them in terms of their positives/negatives and costs/benefits.

Typical mistake

Note that even if carbon emissions were stabilised within the next few years, global warming and climate change would continue for many decades.

Summary

- Carbon and water are vital elements to the survival of life on Earth.
- For both the carbon and water cycles, understand the meaning of the lithosphere, hydrosphere, cryosphere, atmosphere and biosphere, the major stores of carbon and water — their size and geographical distribution.
- The carbon and water cycles operate as closed systems which have links with the atmosphere, oceans, land and biosphere.
- Key processes affect the flows and transfers of both water (evaporation, transpiration, precipitation and cryospheric processes) and carbon (precipitation, photosynthesis, decomposition, weathering, respiration and combustion).
- The cycling of water exists at the global and drainage basin scales. There are a number of common inputs, outputs, stores and flows.
- Physical and human factors affect water and carbon cycles in the tropical rainforest and Arctic tundra.
- Natural and human factors lead to changes in the water and carbon cycles over time.
- Understand the impacts of changes on the water and carbon cycles; these may be economic, social or environmental.
- There are key links between the water and carbon cycles and the atmosphere, oceans, cryosphere and vegetation (biosphere). Being able to link knowledge is a key feature of A-level geography.
- There are global management strategies to protect the water and carbon cycles.

Exam practice

1 Suggest how farming practices can affect carbon stores. [4]
2 Explain **three** criticisms of mapping global water vapour using satellite imagery, as in Figure 2.12. [3]
3 Examine the impact of human activity on the water cycle in the Amazon rainforest. [10]
4 Examine the concepts of positive and negative feedback on the water cycle. [10]
5 Analyse the impact of long-term climate change on the carbon cycle. [10]

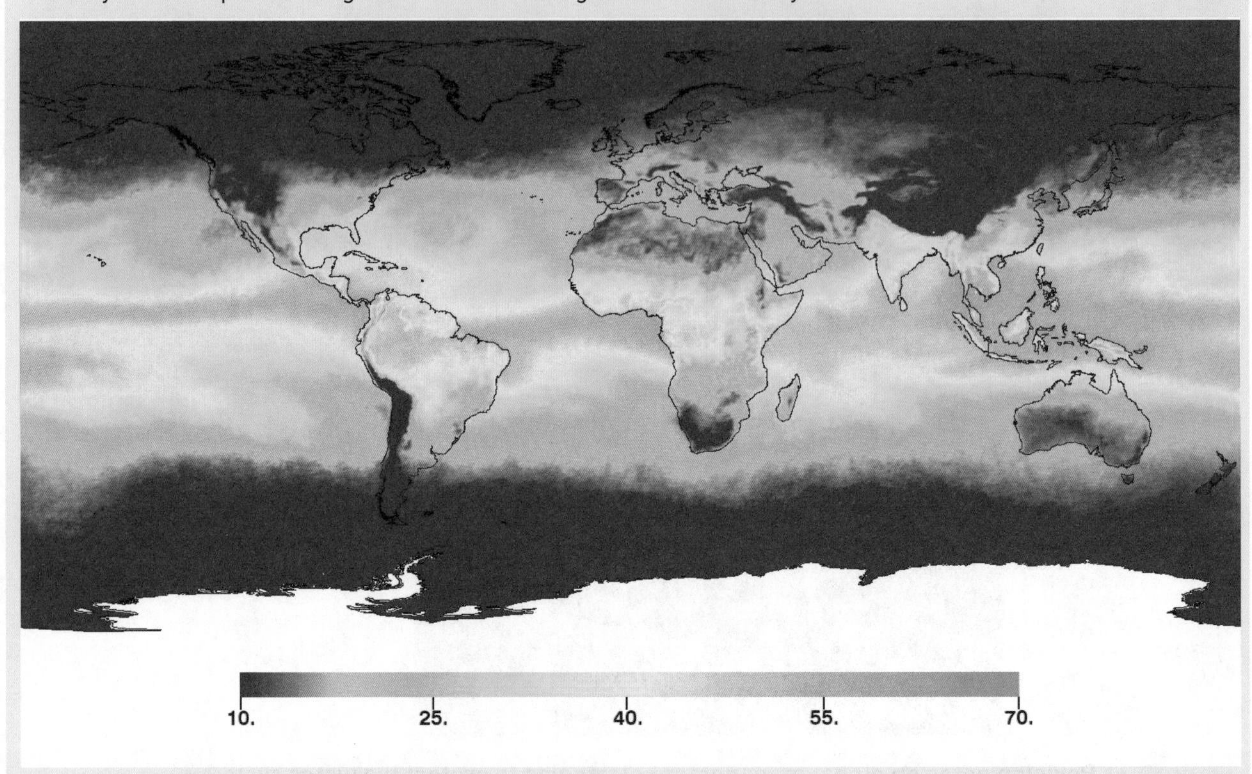

Figure 2.12 Satellite image of global water vapour (mm)

Answers and quick quiz 2 online

ONLINE

3 Changing spaces; making places

What's in a place?

Places are multi-faceted, shaped by shifting flows and connections which change over time

REVISED

Characteristics that contribute to place identity include:

- **physical geography:** e.g. altitude, aspect, drainage
- **demography:** the number of inhabitants, their age, gender and ethnicity
- **socio-economic factors:** e.g. employment types, education
- **cultural factors:** e.g. religion, traditions
- **political factors:** e.g. governance at the local, regional and national level
- **built environment:** e.g. age and style of buildings, building materials
- **history:** e.g. landmarks, historical buildings

Exam tip

Make sure that your local place case studies contain place-specific facts that illustrate contrast — do not just see the two examples in isolation as comparison is important.

Case studies of two contrasting place profiles at a local scale

Revision activity

Use the structure in Figure 3.1 to make a revision summary of the two contrasting local places that you have covered in class.

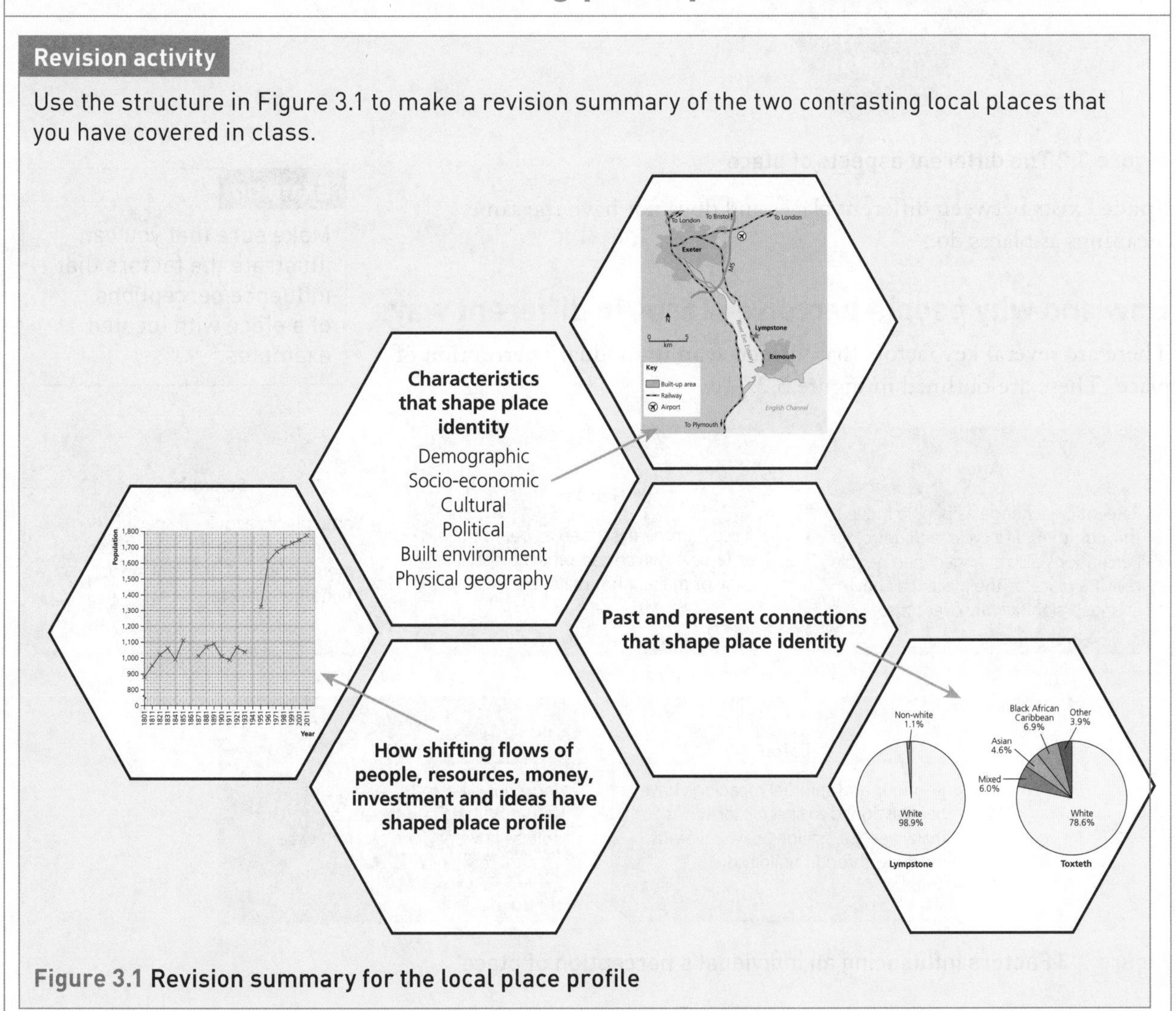

Figure 3.1 Revision summary for the local place profile

How do we understand place?

People see, experience and understand place in different ways; this can also change over time

REVISED

The complexities that exist when trying to define place

Place is a key term in geography. It can be seen as a **location** on a map or more broadly as a **description** of human and physical characteristics. There are three key concepts of place (Figure 3.2):

1 **location:** objective
2 **locale:** objective
3 **sense of place:** subjective

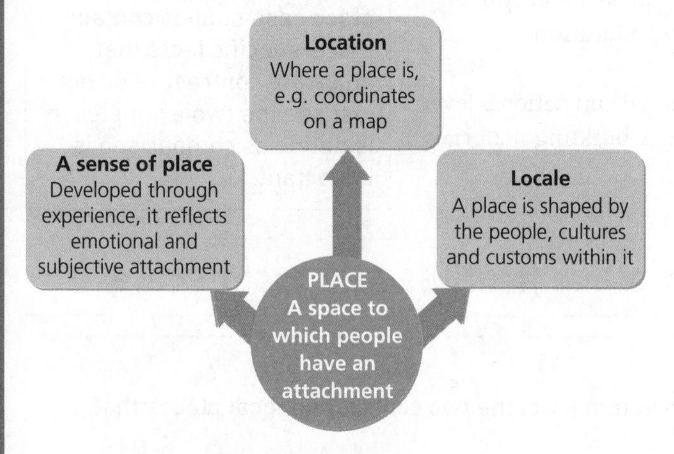

Figure 3.2 The different aspects of place

Space exists between different places and does not have the same meanings as places do.

> **Exam tip**
>
> Make sure that you can illustrate the factors that influence perceptions of a place with located examples.

How and why people perceive places in different ways

There are several key factors that influence an individual's perception of place. These are outlined in Figure 3.3.

Age

Perceptions change as people move through their 'life cycle' and get older. Perceptions also change when people revisit a place as the place may have changed significantly over time.

Gender

Places can be described as being male or female, which can reflect a society's view of male and female roles.

Sexuality

Some places acquire a specific meaning because they are places where people of different sexual orientations cluster.

Religion

Religious and spiritual meanings have been assigned to specific locations for many years. Locations have religious meaning through buildings.

Role

Individuals perform a variety of different roles at different times. Our role at any one time can influence our perceptions of a place.

Figure 3.3 Factors influencing an individual's perception of place

How level of emotional attachment to place can influence people's behaviour and activities in a place

Memory is a very individual and personal thing. There are positive and negative, short- and long-term memories. Memory and personal experience affect how we feel about places. Emotional attachments to places can form through memory and feelings about a place for a variety of reasons: childhood, sporting achievements, milestones in a person's life cycle, national identity. This will in turn affect an individual's behaviour in a specific place.

Revision activity

Summarise an example covered in class which illustrates how emotional attachment to place can influence behaviour and activities.

Now test yourself TESTED

1 What are the **three** key concepts of place?
2 How does space differ from place?
3 What factors influence a person's perception of place?
4 What affects people's behaviour in a place?

Answers on p. 219

How processes of globalisation and time–space compression can influence our sense of place

Globalisation is the growing interdependence of countries through increasing global transactions of goods and services; increasing flows of information, labour and capital; and the widespread transfer of technology. The term 'the global village' has been used to convey the idea of the world having become a 'smaller' place because of time–space compression. Many believe that this has given rise to a new geographical era of '**placelessness**', where global capitalism has eroded local culture and localised identities.

For example, the global spread of retail chains and TNCs, such as McDonalds, Costa Coffee and Hilton Hotels, means that city centres across the world have common elements.

There are different views — some people welcome the impacts of time–space compression, others feel a sense of dislocation. The latter view has given rise to the idea of **glocalisation** — a response to globalisation that centres on the promotion of local goods and services and the adaptation of global products to the specific locality in an effort to regain local cultures and identities.

Now test yourself TESTED

5 How are places being made more similar?
6 How are some places regaining a sense of local identity?

Answers on p. 219

Places are represented through a variety of contrasting formal and informal agencies

REVISED

Informal and formal representations of a place

The way in which places are represented through both formal and informal means can influence how we feel about that place.

- **Informal** ways of representing place: TV, film, music, art, photography, literature, graffiti, blogs.
- **Formal** ways of representing place: census data, statistics, geospatial data, maps.

Exam tip

Develop a critical evaluation of the usefulness of different resources in representing place.

Revision activity

An understanding of the contrast between formal and informal representations of place can best be explored through the use and application of both types of resource. Based on examples studied in class:

1. organise your revision notes into a series of annotated resources. Select a mixture of both formal and informal
2. summarise your key findings regarding the impact and usefulness of the resources by drawing comparisons: e.g. photographs focus on a single image chosen by the photographer; numerical data can have an element of subjectivity and bias

How does economic change influence patterns of social inequality in places?

Resources, wealth and opportunity are not evenly spread within and between places

REVISED

The concept of social inequality and how this is measured

Social inequality is when unequal opportunities or rewards exist for people within a society and between people of different social status or position. When social inequalities lead to substantial differences between groups of people the term deprivation is used. **Multiple deprivation** is the 'lagging behind' of members of society in a number of related aspects of life. It is a social situation that is cyclical in nature and therefore very difficult to break out of (Figure 3.4).

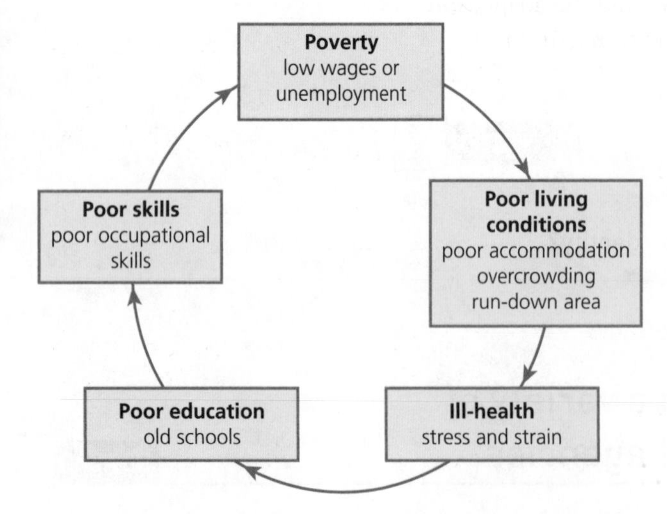

Figure 3.4 The cycle of deprivation

The **Index of Multiple Deprivation** is used by the UK government to spatially assess levels of deprivation. It combines seven factors: income, employment, health, education, crime, access to housing and services, and living environment.

Social inequality can be measured through a variety of data sources.

- **Income — purchasing power parity (PPP):** the World Bank's definition of absolute poverty is US$1.25/day PPP. The use of PPP allows global comparisons to be made between countries as costs vary considerably from one country to another.
- **Housing:** housing tenure is an important indicator of social inequality. Figures are often based on different types of housing tenure. Tenure may be owner-occupied, where people own their house, usually through borrowing in the form of a mortgage; charities and housing associations provide housing; housing can be rented; squatter settlements refer to places where people have no legal right to the land they occupy.
- **Education:** formal education through, for example, schools and colleges, and informal education through skill acquisition. Literacy levels is the most commonly used measure of inequality in education, particularly between countries.
- **Health care:** access to health care and levels of ill-health are closely associated with social inequality. Measures used for the comparison of health care include doctor-to-patient ratios. Within a country, measures such as 'the postcode lottery' can be used — this relates health care provision to where you live.
- **Employment:** this has a direct impact on levels of income, standard of living and quality of life. Wages vary significantly across different forms of employment and **informal employment** is also an important consideration in low-income developing countries (LIDCs) and emerging and developing countries (EDCs). Employment and unemployment figures can often be quite complex and sometimes difficult to access.

Exam tip

Multiple deprivation levels are mapped at the neighbourhood or Lower Super Output Area level. Describing patterns is an important geographical skill.

Typical mistake

Remember that deprivation can also exist in rural areas.

Exam tip

Remember that factors such as access to clean water, sanitation, diet, type of housing and air quality also affect health.

Informal employment is work that is not protected or regulated by the state.

How and why spatial patterns of social inequalities vary within and between places

Inequality exists at all spatial scales from global to local, between urban and rural areas, and within urban areas. Several factors often combine to explain spatial patterns of inequality. These are summarised in Table 3.1.

Table 3.1 Factors explaining spatial variations in social inequality

Factor	Description
Wealth	Wealth influences wellbeing and quality of life; low incomes are linked to factors such as ill-health and poor access to services. Qualifications and skills enable an individual to raise their income level and improve their quality of life. **Disposable income** is an important consideration when discussing wealth
Housing	Income affects housing choices. In LIDCs and EDCs millions have no choice but to occupy slum dwellings. Homelessness is a growing problem in advanced countries (ACs); an additional problem is when house prices rise at a faster rate than wages. This can lead to a shortage of affordable housing at the lower end of the market ➜

Disposable income is the proportion of a person's income that remains after spending on essentials such as taxes, housing and food.

Factor	Description
Health	There is a clear link between deprivation and ill-health. Contributing factors include poor diet, sub-standard housing and stress. Health care varies spatially and access to health care services can be a problem for certain groups in societies, e.g. the elderly, and in certain areas, e.g. remote rural areas
Education	Achieving universal primary education was one of the Millennium Development Goals and many governments invest in education to boost employment and raise living standards. Access to education can be a problem in some areas, e.g. remote rural areas in LIDCs
Access to services	There are wide variations in access to services between countries at varying levels of economic development. Measures include number of doctors per 1,000 people. At a national scale, there are also differences between regions, within regions and between urban and rural areas. All types of services should be considered, not just health services: for example in the UK there is a 'digital divide', reflecting varying levels of access to the internet

Now test yourself

TESTED

7 What is multiple deprivation?
8 How is it measured?
9 State **two** factors that explain spatial variations in social inequality.

Answers on p. 219

Processes of economic change can create opportunities for some while creating and exacerbating social inequality for others

REVISED

The influence of globalisation in driving structural economic change in places

Globalisation has led to increased interconnectedness of the world. This has led to considerable economic change:

- TNCs have a key role in driving economic change and impact the lives of billions of people.
- The **global shift** refers to change in manufacturing areas from western Europe and North America to NICs (newly industrialising countries) in East Asia and Latin America.
- ACs have been transformed into post-industrial societies in which there are high levels of employment in the tertiary and quaternary sectors.

Globalisation is a process leading to the growing integration and interdependence of people's lives.

Deindustrialisation is the reduction of industrial activity in a region or country.

Now test yourself

TESTED

10 How can TNCs impact on people's lives?

Answers on p. 219

Exam tip

Remember that structural economic change also has social and environmental impacts.

The impact of structural economic change on social opportunity and inequality for people and places

Many ACs have undergone a process of **deindustrialisation** when they lost a competitive advantage in many manufacturing sectors. In the 1970s in the UK, cities, economies and societies were transformed by this process. It triggered a **vicious circle of decay**, leading to multiple deprivation in many urban areas:

declining job opportunities → rising unemployment → decline in services → physical environment and infrastructure deteriorates → economically active people move away → declining tax base → increasing decay → loss of investment confidence → declining jobs

This circle of decay is also referred to as the downward **multiplier effect**. A reversal may take place with the positive impacts from investment and job creation — the upward multiplier effect. Table 3.2 summarises the positive and negative impacts of economic change on people and places.

The **multiplier effect** is the process by which a new or expanding economic activity in an area creates additional employment. As employees have money to spend, growth occurs in other sectors and the wealth of the area stimulates more economic activity and investment, creating more jobs and so on. The effect can be positive (upward multiplier) or negative with a downturn in the economy from job losses (downward multiplier).

Table 3.2 Positive and negative impacts of economic change on people and places

	Positive	Negative
In ACs	• Cheaper imports of all relatively labour-intensive products can keep cost of living down and lead to a buoyant retailing sector • Greater efficiency apparent in surviving outlets. This can release labour for higher productivity sectors (this assumes low unemployment) • Growth in LIDCs may lead to a demand for exports from ACs • Promotion of labour market flexibility and efficiency, with greater worker mobility to areas with relative scarcities of labour, should be good for the country • Greater industrial efficiency should lead to development of new technologies and promotion of entrepreneurship, and should attract foreign investment • Loss of mining and manufacturing industries can lead to improved environmental quality	• Rising job exports lead to inevitable job losses. Competition-driven changes in technology add to this • Job losses often affect unskilled workers • Big gaps develop between skilled and unskilled workers who may experience extreme redeployment differences • Employment gains from new efficiencies will only occur if industrialised countries can keep their wage demands down • Job losses are invariably concentrated in certain areas and certain industries. This can lead to deindustrialisation and structural unemployment in certain regions • Branch plants are particularly vulnerable as in times of economic recession they are the first to close, often with large numbers of job losses
In EDCs and LIDCs	• Higher export-generated income promotes export-led growth — thus promoting investment in productive capacity. Potentially leads to a multiplier effect on national economy • Can trickle down to local areas with many new highly paid jobs • Can reduce negative trade balances • Can lead to exposure to new technology, improvement of skills and labour productivity • Employment growth in relatively labour-intensive manufacturing spreads wealth, and does redress global injustice (development gap)	• Unlikely to decrease inequality — as jobs tend to be concentrated in core regions of urban areas. May promote in-migration • Disruptive social impacts, e.g. role of TNCs potentially exploitative and may lead to sweatshops. Also branch plants may move on in LIDCs too, leading to instability (e.g. in the Philippines) • Can lead to overdependence on a narrow economic base • Can destabilise food supplies, as people give up agriculture • Environmental issues associated with over-rapid industrialisation • Health and safety issues because of tax legislation

Figure 4.36 The positive and negative impacts of global shift from Nagle, G. and Guinness, P. (2011) *Geography for Cambridge International A and AS Level*, Hodder, p. 453

How cyclical economic change impacts on social opportunities and inequality

Economies of places at any scale are dynamic not static. Economic booms and recessions impact people and places in a variety of ways. During a recession people's spending power is reduced, they make cutbacks and a range of service and retail activities will be impacted. Job losses can activate the downward multiplier in a region. In a time of economic boom, the reverse will be true and **core regions** will develop a strong, positive/upward multiplier effect.

Now test yourself

TESTED

11 What is deindustrialisation?
12 Why were cities in the UK badly affected by deindustrialisation?
13 Explain the positive/upward multiplier effect.

Answers on p. 219

The role of government in reducing, reinforcing and creating patterns of social inequality in places

Governments operate at both the national, regional and local scale within a country. In the UK there are national, county, city and parish councils. There are also trans-national levels of government, e.g. EU. Social inequality can be governed through the following:

- Taxation, e.g. higher income tax for high earners.
- Subsidies, e.g. free school meals.
- Planning, e.g. upgrading of council housing.
- Education, e.g. training programmes or health initiatives.

Exam tip

Remember that sometimes government action may have unforeseen consequences, which reinforce or widen social inequality.

Revision activity

Choose a named example at any scale (national, regional or local) of a government policy that has (a) reduced, (b) reinforced and (c) created social inequality in a place through spending or cuts. You can summarise your notes in a simple table, such as Table 3.3.

Table 3.3 Consequences of government action

Government policy that has **reduced** patterns of social inequality		
Location	**Policy**	**Impact**
Government policy that has **reinforced** patterns of social inequality		
Location	**Policy**	**Impact**
Government policy that has **created** patterns of social inequality		
Location	**Policy**	**Impact**

Social inequality impacts people and places in different ways

REVISED

Case studies of two contrasting places

Revision activity

Using the example in Figure 3.5 for layout (or another method used in this book), make revision notes on how social inequality has impacted people and places in different ways.

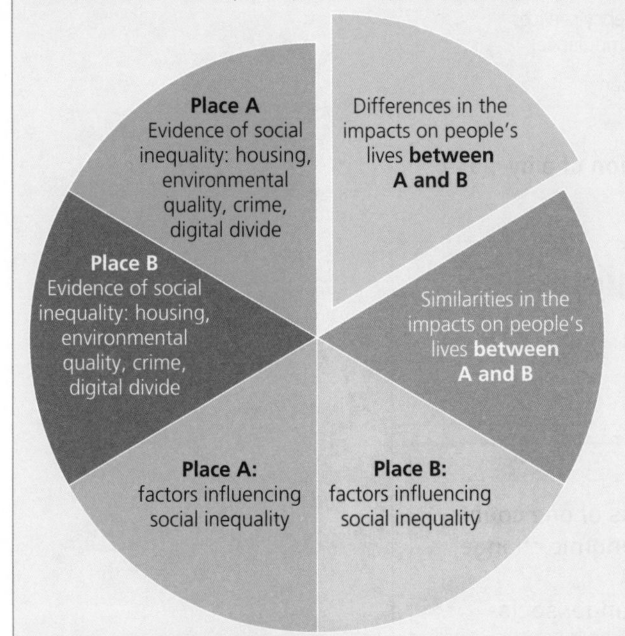

Figure 3.5 Contrasting place examples to show the impacts of social inequality on people and places

Who are the players that influence economic change in places?

Places are influenced by a range of players operating at different scales

REVISED

The role of players in driving economic change

Players are individuals, groups of people or formal organisations who can influence the process of change; some have more influence than others. Another word for players is stakeholders. Public players include government at the national and local scale, while private players include TNCs; there are also non-government organisations to be considered. Figure 3.6 shows an example of the possible players involved in the construction of a by-pass.

Figure 3.6 **Possible players involved in the construction of a by-pass**

Case study of **one** country or region that has been impacted by structural economic change

Revision activity

Make revision notes on an example covered in class of one country or region that has been impacted by structural economic change. Focus on the following headings:

- characteristics of the place before economic change: socio-economic, demographic, cultural and environmental factors
- the economic change/changes that took place and the role of different players in driving change
- impacts on people: socio-economic, demographic, cultural and environmental

How are places created through placemaking processes?

Place is produced in a variety of ways at different scales

REVISED

The concept of placemaking and the role of governments and organisations

Placemaking is an approach to the planning, design and management of public spaces. It is a creative and collaborative process that includes design, development, renewal and regeneration of places. The outcome should be sustainable, well-designed places which meet communities' needs and improve their quality of life.

Placemaking is now an important part of government from national level to local. One example is that of attracting foreign direct investment (FDI). TNCs have considerable choice when identifying suitable locations for investment. Many TNCs invest in ACs, e.g. operations of Barclays and Sony; however, increasingly TNCs from LIDCs and EDCs are spreading their influence regionally and globally. Once they have chosen a location, investment by TNCs will have a significant impact on the placemaking process.

How architects and planners attempt to create meaningful and authentic places

Architecture makes a significant impact on placemaking through the design of individual buildings. Local authorities in the UK develop a Local Plan which has a strategy for new building and developments under the guidance of the Royal Town Planning Institute.

Building design can create positive and/or negative feelings, e.g. negative attitudes towards residential tower blocks built in the 1960s and positive attitudes towards low-density housing areas with green spaces.

How local community groups shape the place they live in

Local communities can have a considerable influence in shaping places through, for example, residents' associations with a concern for housing and the environment, or heritage associations which are active in placemaking that focuses on the protection of historical architecture.

Revision activity

For each of the three examples in this section (governments and organisations, architects and planners, and local community groups), select and make revision notes on a suitable example that illustrates the concept of placemaking.

The placemaking process of rebranding constructs a different place meaning through reimaging and regeneration

REVISED

Why places rebrand to construct a different place meaning

All places have an image that affects people's perception of that place. Many places compete for investment — if a place has a negative image or brand (its popular image) then it may need to rebrand in order to become successful. Rebranding has key elements, which are summarised in Figure 3.7.

Exam tip

Image is a strong determinant of human behaviour. People's perception of place is formed by positive images, particularly when they have no lived experience of that place.

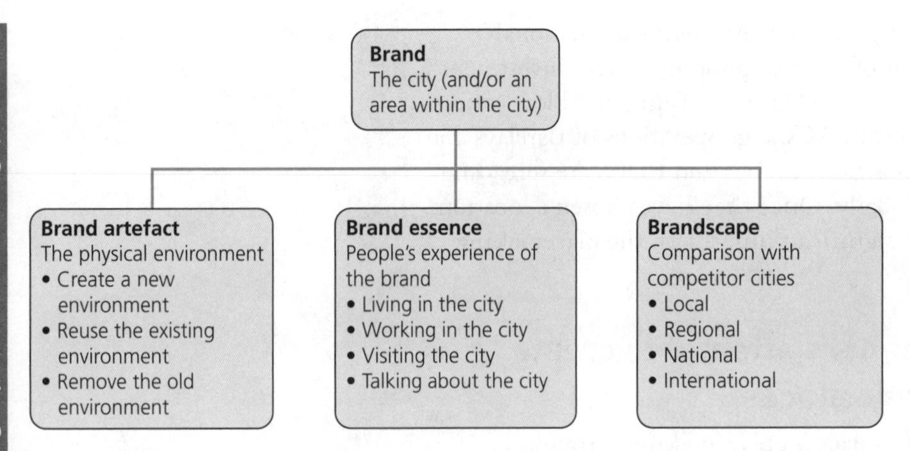

Figure 3.7 Key elements of rebranding

Figure 3.8 shows how rebranding is connected to other attempts to manage or direct the perception of a place.

Figure 3.8 Attempts to manage place perception

How a range of strategies can be used to rebrand places

Types of strategy used to rebrand a place include:

- **market-led:** involving private investors seeking to make a profit
- **top-down:** involves large-scale organisations, e.g. planning departments of local authorities and development agencies
- **flagship development:** large-scale property projects which aim to act as a catalyst for further investment
- **legacy:** following an international or high-profile sporting event that brought investment and regeneration to a place
- **events or themes:** major events that often focus on cultural development

Various distinctive elements are involved in the rebranding process. These include architecture, heritage, retail, art, sport and food.

Typical mistake

Do not see attempts to manage place perception in isolation. Frequently, reimaging and regeneration lead to rebranding.

Now test yourself TESTED ☐

14 Define the following:
 (i) rebranding
 (ii) reimaging
 (iii) regeneration
15 What is top-down rebranding?

Answers on p. 219

A range of players and their role in placemaking

The term players or stakeholders is used to summarise the wide range of people involved in and affected by the process of rebranding. Key players are often those involved in providing the funding for rebranding. These may include:

- governments at various scales (local councils, national governments and regional government organisations such as the EU's European Regional Development Fund)
- corporate bodies (banks, insurance companies and development companies)
- non-profit-making organisations (the National Trust, local community groups)

Exam tip

Advertising and reimaging can make a significant impact on people as they often act on perception rather than objective facts.

Revision activity

Use Table 3.4 to record examples of how different stakeholders (e.g. governments, corporate bodies) have had an impact on placemaking.

Table 3.4 A range of stakeholders/players and their role in placemaking

Stakeholder	Role	Impact

How and why some groups of people contest efforts to rebrand a place

Views on rebranding may include the following.

- **Negative views of local residents because of gentrification.** Gentrification often leads to wealthier people moving into an area. As a result, not only does the socio-economic nature of the local population change but also the provision of services, such as corner shops being replaced by restaurants and wine bars. Local house prices will also increase, forcing some local residents out of the property market.
- **Resentment as one group is favoured by rebranding.** Some retail developments suit a more affluent visitor rather than those living in close proximity where there is less affluence and a need for a different range of facilities, e.g. Liverpool One, a large retail centre in inner-city Liverpool.
- **Negative attitudes regarding the focus of spending.** There will be a variety of opinions regarding the costs and benefits of any rebranding project and new services and facilities provided as part of the project. In times of economic recession, when financial resources are stretched, an expensive retail investment may not be seen as worthwhile at a time when people are cutting back on spending.

Exam tip

Develop a critical evaluation of attempts to manage place perception.

Typical mistake

Note that rebranding can occur as a by-product of regeneration.

Making a successful place requires planning and design

REVISED

Case study of one place that has undergone rebranding

Revision activity

Using your case study example covered in class, make revision notes on a place that has undergone rebranding. Remember to focus on:

- why the place needed to rebrand
- strategy/strategies involved in rebranding
- the role and influence of a range of players (you could draw on Table 3.4 for this)
- how the rebranding has affected people's perception of that place
- the relative success of the rebranding

Summary

- Places are shaped by a range of characteristics, by past and present connections and by shifting flows of people and resources (natural and financial).
- Key concepts of place give rise to a definition.
- People's perceptions of a place vary according to a range of factors including emotional attachment.
- Processes such as globalisation are impacting on our sense of place.
- Place is represented through a wide range of formal and informal methods, each with their own advantages and disadvantages regarding accuracy of representation.
- Economic change can influence patterns of social inequality in places.
- Social inequality impacts people and places in different ways.
- A range of players operate at different scales to influence economic change in places. These include national governments, TNCs and international organisations.
- Placemaking is produced by architects, planners, governments, organisations and communities.
- Sometimes places are redefined by processes such as rebranding, reimaging and regeneration.

Exam practice

1 Explain the difference between a sense of place and place perception. [4]
2 Explain the various aspects of place. [4]
3 Suggest how census data are more useful than photographs in forming people's perceptions of a place. [4]
4 Analyse the different strategies that can be used to rebrand a place. [6]
5 Explain the difference between reimaging and regeneration in the context of place. [6]

Answers and quick quiz 3 online

ONLINE

Global connections options

Option A Trade in the contemporary world

What are the contemporary patterns of world trade?

There are a wide range of sources of data on trade: the World Trade Organization (WTO), the United Nations Conference on Trade and Development (UNCTAD), the International Monetary Fund (IMF) and the Organisation for Economic Cooperation and Development (OECD) are the main ones to know.

International trade involves flows of merchandise, services and capital

REVISED

An understanding of the components of international trade

A geographical feature of the volume and pattern of international trade is its unevenness. Trade is, in the main, dominated by the advanced and rapidly emerging economies which have the economic wealth and political power to negotiate a favourable trade position. The least developed countries have limited access to global markets and a much weaker negotiating position. Figure 3.9 shows the main flows of inter-regional trade.

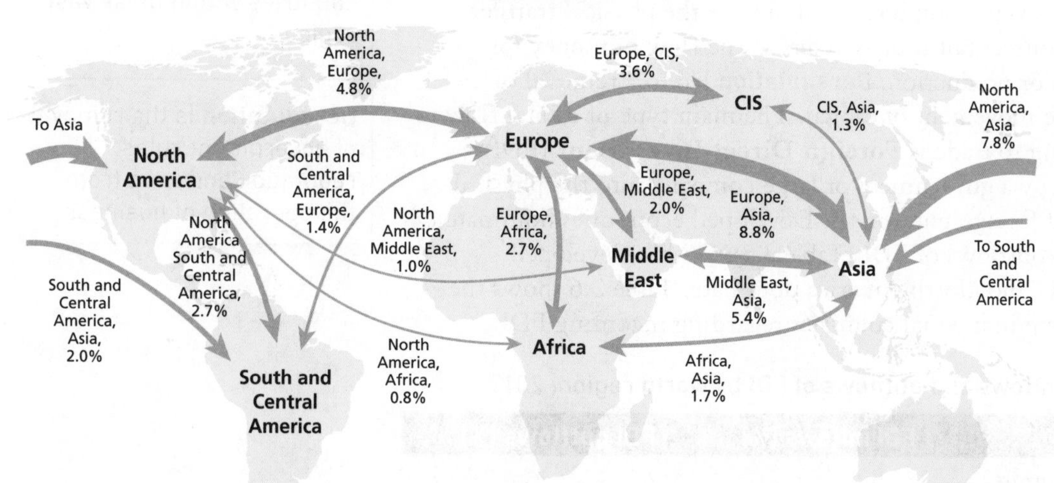

Figure 3.9 Global patterns of inter-regional trade

Merchandise

International trade in merchandise is the inward and outward movement of goods through a country. General patterns across world regions (Figure 3.10) show:

- high values of merchandise exports in the economies of Europe and Asia
- very low values of merchandise exports in Africa
- Asia has nearly ten times the value of merchandise exports of Africa
- Europe has nine times the value of exports of South and Central America
- the export value of fuel and mining products is greatest in the Middle East
- agricultural exports are a strong feature of the EU countries

Exam tip

Be able to identify and name the flows that play a key role in globalisation.

- export of manufactured goods from Asia is 40 times greater than that from Africa

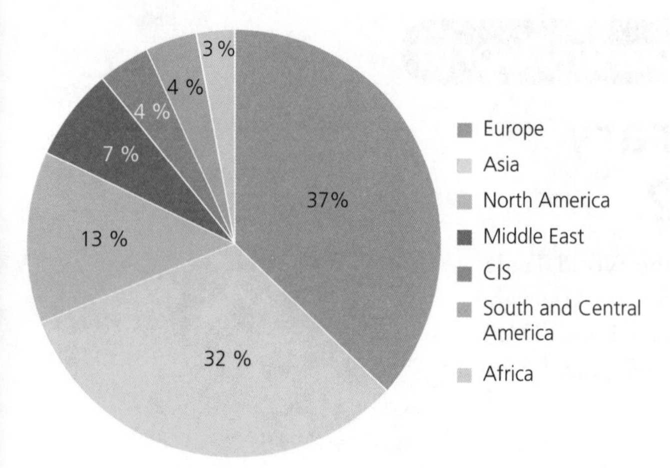

Figure 3.10 The volume and pattern of global merchandise trade, 2013

Services

The global pattern of commercial services exports is also very uneven. Europe is the largest net exporter of commercial services. African countries, especially those in sub-Saharan Africa, are overall net importers of commercial services. The most important exporters of commercial services are the countries of the EU, the emerging economies of Asia, especially India and China, North America and Japan.

Exam tip

Remember that figures representing trade across global regions give a general impression and there will be variations between countries within these vast regions.

Capital

The term capital is very complex: it can involve the physical transfer of production resources but usually refers to the flow of money for investment, trade or production. **Deregulation** led to a removal of restrictions on the movement of capital. The main type of capital flow to understand relating to trade is **Foreign Direct Investment** (FDI) — an investment made by a government or large company into the physical capital or assets of foreign enterprises. Developed economies dominate both inflows and outflows of FDI (Table 3.5). Within developed economies the EU and North America dominate. Table 3.6 shows the dominance of some individual countries regarding incoming FDI.

Deregulation is the removal of government rules, regulation and laws from the operation of business.

Table 3.5 Major inflows and outflows of FDI by world region, 2012

Region	FDI inflows (%)	FDI outflows (%)
Developed economies	41.5	65.4
Developing economies	52.0	30.6
Africa	3.7	1.0
Asia	30.1 (of which E & SE Asia 24.1)	22.2 (of which E & SE Asia 19.8)
Latin America & Caribbean	18.1	7.4
Oceania	0.2	0.0
Transitional economies	6.5	4.0
Structurally weak, small, vulnerable economies	4.4	0.7

Source: UNCTAD statistics

Table 3.6 Incoming FDI in the top ten countries, 2011

Country	Incoming FDI US$ billion
USA	258
China	220
Belgium	102
Hong Kong	90
Brazil	72
Australia	66
Singapore	64
Russia	53
France	45
Canada	40

Source: UNCTAD

- Flows of capital in the global trade system are complex as capital includes many diverse elements and the growing interconnectivity of trade has increased global capital flows.
- Most investment flows are intra-firm, especially within TNCs.
- FDI is a major source of global investment.
- FDI is an important source of funding for development in all countries, but particularly in countries with plentiful natural resources (e.g. Brazil), large consumer markets (e.g. China) or growing financial services (e.g. Hong Kong).

Now test yourself

TESTED

16 Using Figure 3.9, which regions are fully connected to the global patterns of international trade and which are under-represented?
17 Describe and account for the global trends in merchandise trade.
18 Suggest why the inflow of FDI to developing economies is much greater than the outflow of FDI.

Answers on pp. 219–20

Current spatial patterns of inter-regional and intra-regional trade

Inter-regional trade between Europe and North America

The EU and USA are long-established trading partners now accounting for over 30% of global trade. The main exports from the USA to the UK are aircraft, machines, gems/precious metals and coins. The main UK exports to the USA are machines, vehicles and oil.

Intra-regional trade within the EU

The intra-regional trade within the EU is complex due to 28 (currently) member states all with a diverse range of products and services in which they have a **comparative advantage**.

Exam tip

Make sure that your facts regarding trading relationships are up to date. For example, the political events of 2016 (the Brexit vote and the US presidential election) will redefine key trading relationships.

Comparative advantage: the principle that countries or regions benefit from specialising in an economic activity in which they are relatively more efficient or skilled.

Revision activity

Using an example studied in class, make notes on the **factors** that could influence the pattern of intra-regional trade, for example between a single country and its exports to EU member states of one product. You could organise the notes into a table or diagram. Remember to consider economic, political, social and environmental factors.

Exam tip

Good sources for up-to-date facts include the following websites:
www.worldbank.org
www.unctad.org
www.wto.org

Current patterns of world trade are related to global patterns of socio-economic development

REVISED

The relationship between patterns of international trade and socio-economic development

There is a strong positive relationship between trade and development which can be shown through a comparison of Human Development Index figures and export values.

Revision activity

Select two countries at different stages of the development continuum and examine their links between trade and socio-economic development.

How international trade can promote stability, growth and development

International trade has benefits for all countries, it can also help developing countries progress and become integrated in the global trading system. The WTO recognises such benefits and works to support such efforts. Examples of how trade can promote stability, growth and development include:

Stability

- International trade can reduce price fluctuations and equalise prices of goods.
- Trade promotes co-operation and understanding between nations.
- Trade agreements can lead to political stability in the countries involved.

Growth

- International trade means that some goods are produced at a large scale in order to satisfy demand, the economies of scale achieved have advantages for all in lower prices.
- Trade leads to optimum use of natural resources.
- Trade leads to specialisation.
- Global competition for goods produced promotes efficiency.

Development

- Trade means that a wide range of goods can be made available through imports.
- International trade promotes an exchange of technical knowledge and the establishment of new industries.
- Trade requires the development of transport networks.
- Trade raises living standards through the economic multiplier effect.

How international trade causes inequalities, conflict and injustices

The global pattern of trade is uneven. Unequal flows of money, ideas, people and technology can lead to inequalities, conflict and injustices. The main flows of people are from poor to rich areas leaving countries with skills shortages. Developing countries often lag behind in technological advances and investment by MNCs can lead to widening inequality in host countries. Examples of how trade can cause inequality, conflict and injustices include:

Inequalities

- There can be limited access to trading blocs for some countries.
- Some groups of society benefit more from the advantages of trade depending on skills set and gender.
- Some places benefit more from the advantages of trade – theory of growth poles and core/periphery models.

Conflict

- International trade can threaten 'home industries' leading to conflict.
- Trade disputes over wages and conditions.

Injustices

- LIDCs can be economically exploited by ACs.
- Sometimes goods are exported to earn foreign exchange but are in short supply within a country.
- Practices such as land grabbing for agricultural production can create injustices for local farmers and threaten food security.

Now test yourself

TESTED

19 Explain the concept of comparative advantage.
20 How does international trade promote economic growth?
21 Name **two** ways in which international trade can
   (a) cause conflicts
   (b) lead to injustices

Answers on p. 220

Why has trade become increasingly complex?

Access to markets is influenced by a multitude of inter-related factors

REVISED

International trade has increased connectivity due to changes in the 21st century

Technology, transport and communications

Technology

Global supply chains represent the flows of materials, products, information, services and finance through a global network. This has increased connectivity within the global trade system. The World Economic Forum and the OECD recognise several 'pillars' which ease integration, e.g. ICT, transport investment and information systems.
It is now common for different stages of the production process to be located in different parts of the world through global supply chains. The global production network for any company involves a complex system of flows of information, finance, components and finished goods. These international production networks are organised differently for different industries and as costs vary. **Just-in-time** (**JIT**) technology has also enabled a range of cost-saving advantages in the production of goods, e.g. part-assembled goods ready to be quickly finished and distributed as and when required. Industries using JIT technology include car manufacture and the computer industry.

Just-in-time (JIT) is a system companies use to increase efficiency and decrease storage costs and waste. Products are delivered as they are required.

Exam tip

India's development as a global service centre shows the influence of telecommunication technology and the decreasing significance of distance.

Transport

The increased size and standardisation of many containers used to transport goods have meant that transport of manufactured goods is now more efficient. Reductions in cost, computerised logistics systems and data analysis of efficiency of handling and distribution have also facilitated freer movement, e.g. the use of standardised containers for sea, rail, road and air transport. Investment in these new developments can be very costly and this can disadvantage developing countries.

Communications

Advances in technology are one of the main reasons for rapid globalisation. In information and communication technology, innovations have become smaller in size, more efficient and often more affordable to both individuals and businesses. Digital connectivity is vital for producers and customers in the delivery of services such as finance. ICT is also central to border security. Quality of ICT and broadband access varies with location at all scales but particularly significant is the limited access in developing countries which can hinder trade and development.

Increasing influence of MNCs in EDCs

International trade is dominated by multinational corporations (MNCs), the top 500 of which account for 70% of the flow of goods, services and capital. Most have headquarters in ACs with branch plants in EDCs. Investment by an MNC brings both costs and benefits for the host nation, as outlined in Table 3.7.

Table 3.7 The impact of MNC operations for host countries

	For the host country
Benefits	● Generates jobs and income ● Brings new technology ● Gives workers new skills ● Has a multiplier effect ● Raises living standards ● Improves trade balance
Problems	● Poor working conditions ● Exploitation of resources ● Negative impacts on environment and local culture ● Economic leakages/repatriation of profits ● Reduces economic security ● Possible closure of 'home' industries

Outsourcing is a cost-saving strategy where a company that has a comparative advantage provides goods or services for another company even though they could be produced in-house. In 2015 European businesses outsourced significant amounts of services, e.g. UK 17%. India is a main destination for the outsourcing of contracts in IT services.

Typical mistake

MNCs are not confined to the manufacturing sector. Many well-known companies in the service sector are MNCs, e.g. HSBC and Starbucks.

The role of regional trading blocs

The main global trading blocs, which allow free trade between group members, are NAFTA (North American Free Trade Agreement: Canada, Mexico and USA), the EU (European Union: 28 member states at present, pending Brexit) and ASEAN (Asian Free Trade Area: 10 member states).

Advantages of trade agreements:

- economic development, through the **economic multiplier**
- intergovernmental support and security
- allow representation and influence in world affairs
- freedom of movement of goods and labour
- easier negotiation of trade with other large trading groups
- sharing of technological advances

Disadvantages of trade agreements:

- lack of access to trading blocs by poorer nations can widen the **development gap**
- trade disputes can arise over tariffs, prices of commodities and changes in trade agreements
- border and customs authorities can be subject to corruption and breaches of security
- some loss of sovereignty
- pressure to adopt central legislation

Economic multiplier: the process whereby expanding economic activity creates additional employment leading to further growth and investment.

Development gap: the difference in prosperity and wellbeing between rich and poor countries.

The main global trading blocs have strong negotiating powers and account for most global trade. The EU is the largest economic and political union and in 2013 was responsible for over 16% of world trade.

In 2013, it traded with 59 countries. The EU operates a single market allowing the free flow of goods within the EU.

The growth of 'South–South' trade

World trade between LIDCs and EDCs is increasing and linking the growing markets in Asia, Africa and Latin America. In 2013 South–South trade accounted for 50% of China's trade and 60% of trade in India. Reasons for the increase include:

- rising demand in India and China for raw materials and energy
- a growing market potential in Asia and Latin America
- increasing demand from the growing middle class in Brazil, China and India
- growing FDI from China and India into developing countries in Latin America and Africa

The growth of services in the global economy

Services is the growth sector in international trade across all world regions. Europe is the highest exporter but Asia's share is growing rapidly. Travel and tourism have accounted for growth in developing countries, e.g. Cambodia and Uganda. Growth in services is important for development in EDCs and LIDCs.

Increasing labour mobility and the new international division of labour

Labour is the human resource available in the economy. It is less mobile than capital but there have been significant increases in global economic migration. Although the pattern is constantly changing, in broad terms the main flows are from South Asia, Africa and Latin America to North America, Europe and the Gulf countries in western Asia.

Most movements are within geographical regions and between neighbouring countries. Most migrants travelling longer distances are those with education and financial means. The **new international division of labour** refers to the global reorganisation of production due to (a) **deindustrialisation** in ACs and (b) the spread of MNCs. This has produced a pattern of higher-paid jobs in ACs (managerial and research and development) and lower-paid jobs (manufacturing) in LIDCs.

Revision activity

Produce a flow diagram to explain how growth of trade in services can enhance the development process in LIDCs and EDCs. Think about the multiplier effect and how funds can be invested.

Deindustrialisation is the absolute or relative decline in manufacturing in the economy of a country or region.

Now test yourself

TESTED

22 How have transport developments increased international trade?
23 What is outsourcing?
24 What is the new international division of labour?
25 How does it affect global employment patterns?

Answers on p. 220

There is interdependence between countries and their trading partners

REVISED

Case study of one EDC

An example of how you could set out revision notes for an EDC is shown in the case study of India, which you will find online (see p. 3). Remember the focus is:

- direction and components of current international trade patterns
- change in international trade patterns over time
- economic, political, social and environmental interdependence with trading partners
- impacts of trade on the EDC including economic development, political stability and social equality

Exam tip

The multiplier effect is a broad concept which needs a full explanation in an exam answer.

What are the issues associated with unequal flows of international trade?

International trade creates opportunities and challenges which reflect unequal power relations between countries

REVISED

Exam tip

When referring to case studies do not forget to quote specific places, patterns, trends and statistics. Use colour to make these facts stand out in your revision notes.

Case study of one AC to show how core economies have strong influence and drive change in the global trade system to their own advantage

Revision activity

Make your own revision notes on a case study covered in class for this part of the course. Remember to illustrate through economic, political and social factors to explain:

- the country's advantages for trade
- opportunities
- challenges

Case study of one LIDC to show how peripheral economies exert limited influence and can only respond to change in the global trade system

Revision activity

Make your own revision notes on a case study covered in class for this part of the course. Remember to illustrate through economic, political and social factors to explain:

- trade components
- why there is limited access to global markets
- opportunities
- challenges

Summary

- International trade involves flows of merchandise, services and capital. These flows vary across time and space.
- There is a strong positive relationship between international trade and socio-economic development which can be illustrated through the use of statistics such as the Human Development Index and through examples.
- Twenty-first century developments in technology, transport, communications and through MNCs and trading blocs have transformed patterns of international trade.
- A case study of an EDC can be used to illustrate the interdependence between countries and trading partners.
- The opportunities and challenges created by international trade vary between core and peripheral countries.

Exam practice

6 Suggest **two** reasons for the growth of 'South–South' trade. [2]

7 Explain the importance of incoming foreign direct investment (FDI) to a country's economy. [3]

8 With reference to a case study, explain the impacts of trade on an economically developing country (EDC). [8]

9 With reference to a case study, explain how access to global markets creates opportunities for low-income developing countries (LIDCs). [8]

Answers and quick quiz 3A online

ONLINE

Option B Global migration

What are the contemporary patterns of global migration?

Remittances: funds sent to a migrant's country of origin.

A **refugee** is a person who has moved outside their country of nationality because of a genuine fear of persecution or death.

An **asylum seeker** is a person who seeks entry to another country by claiming to be a refugee.

Migration is linked to the process of globalisation. The volumes, scale, direction and demographics of migration flows are constantly changing. There exists a wide range of reasons for migration but globally the largest group of migrants are economic migrants seeking work in another country and sending **remittances** to their family back home. There are also a growing number of **refugees** and **asylum seekers**. The impacts of migration on both host country and country of origin can be wide-ranging, and demographic, economic, social, cultural, political and environmental in nature.

Global migration involves dynamic flows of people between countries, regions and continents

REVISED

Current spatial patterns in the numbers, composition and direction of international migrant flows

Typical mistake

Migration does not include such short-term movements as tourism.

Migration is part of the calculation of population change. Net migration is the difference between the number of immigrants and emigrants for a particular country. The main current migration flows are shown in Figure 3.11.

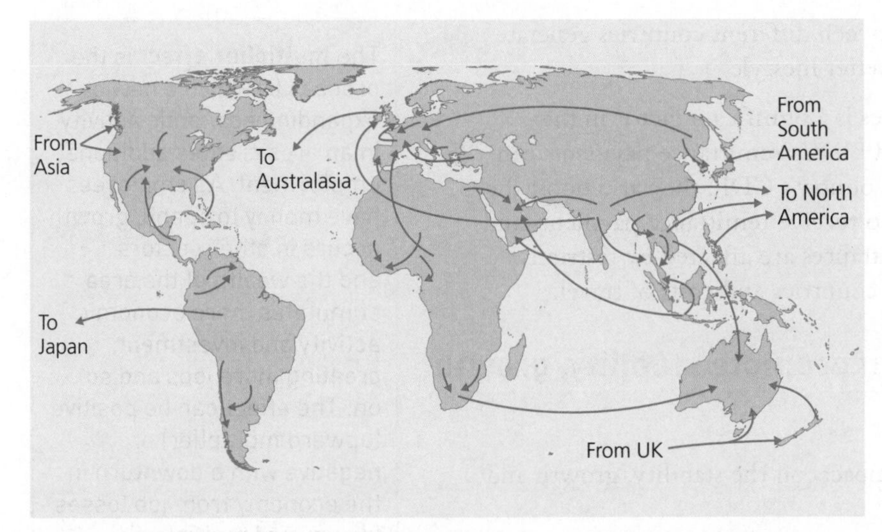

Figure 3.11 Major global migrant flows in the twenty-first century

Table 3.8 summarises examples of international, inter-regional and intra-regional migrant flows.

Revision activity

Annotate a copy of Figure 3.11 with the reasons for **two** of the migrant flows shown.

Exam tip

These spatial classifications of migration could, in the right context, form a useful framework for exam answers.

Table 3.8 Current migrant flows at different spatial scales

International migration	Inter-regional migration	Intra-regional migration
UK international migration • The main countries of origin of UK immigrants are India, Poland and Pakistan • The main countries of destination of UK emigrants are Australia, the USA and Canada • The main reasons for emigration include employment opportunities, retirement and being reunited with families abroad • Women represented just over half of the UK's immigrant population in 2013 • London has 36.2% of immigrants in its population — mostly from India	**Migration to Europe** to escape conflict in Africa and the Middle East • 3,279 people died in Mediterranean Sea crossings in 2014 • Syria and Eritrea accounted for the largest number of migrants arriving in Italy by sea in 2014 • Other routes include crossings between Turkey and Greece and West Africa to Spain	**Migration within the EU** • In 2012, 1.7 million people resident in an EU country migrated to another EU country • Germany is a popular destination for migrants within the EU • The main motive for migration is economic, e.g. Polish migrants attracted to the UK by greater employment opportunities, higher wages and a better standard of living • In 2013 there were 660,000 Poles living in the UK and 40,000 UK residents living in Poland — mostly professionals working for MNCs and attracted by the lower cost of living

Current patterns of international migration are related to global patterns of socio-economic development

REVISED

The relationship between the patterns of international migration and socio-economic development

There is a close relationship between migration and development for two main reasons:

1 Migration can contribute to economic development by providing workforce and new skills; also paid employees lead to the **multiplier effect** (see p.104) as their income stimulates further economic growth.

Exam tip

Remember that, for some countries, receiving immigrants is a selective process in terms of education and skills that can boost the economy and fill specific skills shortages.

2 Inequalities in income levels between different countries generate movement of people seeking a better lifestyle.

In addition, the value of remittances is a significant factor in the development process. In EDCs and LIDCs remittance payments can contribute a large proportion of a country's GDP; they also boost the spending power of the families who receive remittance payments and stimulate economic growth. Remittances are affected by restrictive immigration policies of developed countries and costs of travel.

The **multiplier effect** is the process by which a new or expanding economic activity in an area creates additional employment. As employees have money to spend, growth occurs in other sectors and the wealth of the area stimulates more economic activity and investment, creating more jobs and so on. The effect can be positive (upward multiplier) or negative with a downturn in the economy from job losses (downward multiplier).

How global migration can promote stability, growth and development

Migration can have a number of impacts on the stability, growth and development of a country.

Stability

- Youthful migrants can balance the population structure of a previously ageing population in a host country.
- Countries can become more economically stable as a result of migration, for a host country if the wages are spent within that country the revenue will boost the economy and for a home country, remittances are a source of foreign exchange which will boost the economy e.g. the remittance corridors between the US and Mexico, India and the Philippines.
- Migrants may return home with new ideas of democracy and equality which can contribute to a country's political stability.

Growth

- Migrants in reproductive age groups means an increase in birth rates and population growth. This can have benefits to a host country as a more youthful, expanding population means more workers in the future but also more childhood dependents and more pressure on resources.
- Economic growth can be achieved in a number of ways for the host country; labour/skills shortages can be addressed, migrants may provide cheaper labour, an increase in the size of the workforce can provide an economic boom and trigger the multiplier effect, tax revenue increase. For the country of origin, remittance payments and situations where migrants return home with new skills can both create to economic growth.

Development

- For a country of origin, the reduced pressure on food, water and energy resources, reduced pressure on health services and decline in unemployment can allow development to take place but it is dependent on the demographics of the migrant groups. The situation is helped if remittances are sent home.
- Organisations such as the UN have 'migration and development projects' which partner countries providing and receiving migrants in socio-economic development projects.

- Development can take place when returning migrants bring home new skills sets and are able to maintain linkages which benefit the flow of ideas, resources and finance.

How global migration causes inequalities, conflicts and injustices

Migration can also cause inequalities, conflicts and injustices for both the host country and the country of origin. Conflict can develop for a number of reasons between the migrant population and the host communities, migrants can be exploited through pay and conditions, in the host country they can represent a vulnerable group in society.

Revision activity

Make your own revision summary in a suitable format – diagram, table, bulleted list, of the ways in which migration can cause **inequality**, **conflict** and **injustices**. Remember that both the host country and the country of destination are affected. Include some named examples.

Now test yourself TESTED

26 What is net migration?
27 What are remittances?
28 Why are remittance payments a key element of economic growth in EDCs and LIDCs?
29 State **three** further ways in which global migration can promote economic growth.

Answers on p. 220

Why has migration become increasingly complex?

Global migration patterns are influenced by a range of factors

REVISED

Changes in the 21st century have increased the complexity of global migration

Economic globalisation

Globalisation is leading to the emergence of new source areas and host destinations. Examples exist at three spacial scales:

1 **Inter-regional:** migration of highly skilled workers (graduates, technology and science experts) from China, India and Brazil to the USA. Also, migration of workers from countries such as India, Pakistan and Bangladesh to the oil-producing Gulf States, e.g. Saudi Arabia, attracted by a high demand for labour and a free flow of remittances.

Exam tip

It is useful to link international migration to the globalisation of the world economy.

2 **Intra-regional:** an increase of the **international migrant stock** among ASEAN member states with flows of low-skilled workers from countries such as Cambodia and Myanmar to the fast-growing economies of Thailand and Malaysia.
Increased migration within South America to countries such as Argentina and Chile which can offer more employment opportunities and higher wages than the source areas.
Return migration flows within the EU as young workers who took low-skilled jobs in another country return home with a higher level of skills to more prestigious jobs.
3 **Internal:** internal migration flows, for example within EDCs such as India and Brazil where economic development is concentrated in core areas, often large urban centres.

International migrant stock: the number of people born in a country other than that in which they now live. This includes refugees.

High concentrations of young workers and female migrants

Two significant trends in global migration have been the focus on young age groups seeking the economic benefits of migration and the large percentage of female migrants seeking greater independence. Young workers who are mobile and skilled seek job opportunities and a better quality of life in more affluent countries, e.g. young male construction workers migrating from India to the Middle East and Dubai.

Female migrants make up large percentages of all migrants in regions such as Europe and North America. Reasons include greater independence, women with highly developed skills sets migrating from continents such as Africa and women moving from countries where they still face job discrimination, e.g. China.

Flows of migrants in South–South corridors

In 2013, the South–South international migration flows were greater than the South–North flows.

Reasons include: labour migration, the flow of refugees escaping conflicts in places like Myanmar and South Sudan and restrictive administration barriers for migrants from the South to the North; a number of fast-growing economies in the South; an increase in the awareness of opportunities in countries in the South; and the increased costs of moving long distances from the South to the North.

Conflict and persecution have increased numbers of refugees

A **refugee** is someone who leaves their country of origin or usual domicile because of a fear of persecution or death. According to the UN the number of refugees has risen from 15.7 million in 2012 to 19.5 million in 2014. Syria has become the largest source of refugees, while Turkey was the largest recipient of refugees.

An **asylum seeker** is a person who seeks entry to another country by claiming to be a refugee. Those judged not to be refugees can be sent back to their home country. The largest number of applications are received in the Russian Federation, Germany and the USA.

The main reasons for the large numbers of refugees globally include the effects of conflict (reduced personal safety and destruction of property), political persecution, economic hardship and the impacts of natural hazards.

Typical mistake

Do not think of all migrants as permanent. The aim of a large proportion of migrants is to return home one day.

Revision activity

Make summary revision notes on a located example of a South–South migration corridor covered in class. Remember the importance of place-specific facts.

Changes in national immigration and emigration policies

Migration policies are put in place by individual countries to meet their specific requirements and political needs. The motivations may differ. In ACs, there may be more protectionist measures and migrants may only be accepted if they satisfy a certain skills requirement within the host country, e.g. the points system in Australia. Developing countries will benefit from higher levels of migration where remittances are sent home to form a stimulus to economic growth and a boost to GDP.

Exam tip

Remember the importance of being able to illustrate each of these factors with named examples.

Development of distinct corridors of bilateral flows

A bilateral migration flow is a movement of migrants between two countries. One of the largest and most long-standing is that between the USA and Mexico. Influences on bilateral flows include: close proximity reducing the cost of travel, ease of access, ease of sending remittances, employment opportunities and higher wages, the push effects of conflict or persecution, cultural or historic factors such as language and former colonial influence.

Now test yourself TESTED

30 Why do women make up a large percentage of migrants?
31 State **two** reasons for the increase in South–South migration corridors.
32 Why are the global numbers of refugees increasing?

Answers on p. 220

Corridors of migrant flows create interdependence between countries

REVISED

International migration is an important part of the globalisation process. As with many other aspects of globalisation, some have benefited from increased wealth and opportunities and others have been marginalised. These issues can be illustrated through a case study of an EDC.

Case study of one EDC

Online (see p. 3) you will find a case study of Brazil, illustrating an EDC. This will show how you could set out your revision notes for the case study covered in this part of the course. Remember the revision focus is on:

- current patterns of immigration and emigration
- changes in immigration and emigration over time
- interdependence between countries connected to the EDC by migrant flows
- the impacts of migration

Exam tip

Remember in any discussion of the impact of migrants on a host community to keep a balanced view.

What are the issues associated with unequal flows of global migration?

Global migration creates opportunities and challenges which reflect the unequal power relations between countries

REVISED

Case study of one AC to show how it influences and drives changes in the global migration system

Revision activity

Make your own revision notes on a case study covered in class for this part of the course. Remember to focus on:

- patterns of emigration and immigration, migration policies and interdependence with countries linked by migration
- opportunities
- challenges

Exam tip

Make sure you are familiar with the format of population pyramids.

Case study of one LIDC to show how it has limited influence over and restricted response to the global migration system

Revision activity

Make your own revision notes on a case study covered in class for this part of the course. Remember to focus on:

- patterns of emigration and immigration, migration policies and interdependence with countries linked by migration
- opportunities
- challenges

Summary

- Global migration results in shifts and flows of people over time and place.
- There is a strong relationship between socio-economic development and international migration.
- Global migration leads to costs (inequality, conflict and injustices) and benefits (stability, growth and development) for countries.
- Certain factors have meant that international migration is now extremely complex and ever-changing. These factors include globalisation, conflict and changing policies.
- Study of an EDC can illustrate how corridors of migrant flows create interdependence between countries.
- Case studies of an AC and an LIDC can be used to illustrate how global migration creates opportunities and challenges for countries that drive international migration and those that have limited influence.

Exam tip

Keep case studies relevant and up-to-date with specific details that locate migration flows in time and place.

Exam practice

10 Suggest **two** reasons why females make up a significant percentage of all migrants. [2]
11 Explain how global migration patterns have been influenced by economic globalisation. [3]
12 With reference to a case study, explain the impacts of migration on EDCs. [8]
13 With reference to a case study, explain how emigration from low-income developing countries can provide challenges for that country. [8]

Answers and quick quiz 3B online

ONLINE

Option C Human rights

What is meant by human rights?

There is global variation in human rights norms

REVISED

What is meant by human rights

Human rights are the basic rights and freedoms to which all human beings are entitled. They should protect all individuals, at all times, in all places. The Universal Declaration of Human Rights (UDHR) was adopted by the UN in 1948. It is now evident that 'violations' of human rights have occurred in a range of locations and at a variety of scales. Globalisation has impacted human rights, as has development.

Understanding of norms, intervention and geopolitics

Human rights norms

- Human rights norms are the foundation of human rights.
- There are 30 statements in the UDHR which are accepted as human rights norms.
- Human rights are protected by law and through the signing of international treaties or conventions, e.g. the UN Convention on the Rights of the Child.

Intervention

- Humanitarian intervention (intervention by a state or group of states in a foreign territory) may be used to end human rights violations.
- The UN Security Council is the only body that can legally authorise the use of force.
- There are costs and benefits to such intervention, e.g. political stability or leading to further injustices.
- UN involvement can take many forms, e.g. peacekeeping and the coordination of organisations active in an area.
- Other forms of intervention include economic sanctions and prosecutions of individuals responsible for human rights violations.

Typical mistake

Geopolitical conflicts are often initiated by disputes over natural resources, e.g. oil and water, a cause that is often missed.

Geopolitics

- Geopolitics refers to global political power and international relations. Political power is often closely related to economic power. The USA remains the only superpower.
- The IMF states that there are powerful ACs, increasingly influential EDCs and peripheral LIDCs.
- Organisations such as the UN exert geopolitical influence.
- MNCs have power and influence over the countries in which they invest.

Patterns of human rights violations are influenced by a range of factors

REVISED

Spatial patterns of human rights issues and the factors affecting them

Article 3 of the Universal Declaration of Human Rights (UDHR) states that everyone has a right to 'life, liberty and security'. Forced labour, maternal mortality rates and capital punishment show significant unevenness in their distribution. Figure 3.12 summarises the spatial patterns and factors affecting them.

Forced labour	Maternal mortality rate (MMR)	Capital punishment
• Categories include children forced to work, men unable to leave work because of debts and females exploited • Globally there are 21 million victims; no region is unaffected • Southeast Asia has the highest level at 11.7 million (2012) • Economic factors affecting rates include poverty, migration and low wages • Political factors of influence include conflict, corruption and prejudice • Social factors of influence include gender inequality, sexual exploitation and bonded labour • Environmental factors of influence include escaping climate-related disasters and hazardous working conditions	• Globally in 2013, 289,000 women died during and following pregnancy and childbirth • Most of these deaths occurred in developing countries, e.g. Sierra Leone and Chad • MMR is affected by access to treatments, poor quality medical care, lack of availability of information and education, poverty and cultural barriers • Most of these deaths are preventable and a matter of human rights protected by, for example, the Convention on the Elimination of All Forms of Discrimination Against Women	• According to Amnesty International, in 2014 there were at least 607 executions globally and 2,466 people were sentenced to death in 55 countries • Factors affecting capital punishment include: differences in types of crime for which it is imposed, an increase in the number of countries in which it is being abolished, reinstatement in some countries, changes in the number of pardons

Figure 3.12 Current spatial patterns of human rights issues and the factors affecting them

Now test yourself

TESTED

33 State **two** forms of intervention in human rights issues.
34 What is 'geopolitics'?
35 State **one** social factor that can contribute to vulnerability to forced labour.

Answers on p. 220

Revision activity

Adapt Figure 3.12 to show economic, political, social and environmental factors affecting forced labour, maternal mortality rate and capital punishment. Either use colour coding or add additional sub-sections to the table.

What are the variations in women's rights?

The geography of gender inequality is complex and contested

REVISED

Factors explaining variation in the patterns of gender inequality

Gender inequality is the unequal treatment of individuals based on their gender. Measurements show that it mainly affects women, but increasingly men and boys are included in programmes to tackle gender inequality. The Global Gender Gap Index devised by the World Economic Forum can be used to compare different countries (Figure 3.13): Several challenges show the complexity and contested nature of the issue: these include forced marriages, human trafficking, access to education and violence against women. Factors which explain the variation in the patterns of gender equality are shown in Figure 3.14.

Exam tip

The division of factors into economic, political, social and environmental is important and in the right context provides a useful framework for answers.

Economic participation and opportunity	Educational attainment
1. Labour force participation 2. Wage equality for similar work 3. Estimated earned income 4. Legislators, senior officials and managers 5. Professional and technical workers	1. Literacy rate 2. Enrolment in primary education 3. Enrolment in secondary education 4. Enrolment in tertiary education
Health and survival	**Political empowerment**
1. Sex ratio at birth (female/male) 2. Healthy life expectancy	1. Women in parliament 2. Women in ministerial positions 3. Years with female head of state (last 50)

Figure 3.13 The indices used by the WEF to create the Global Gender Gap

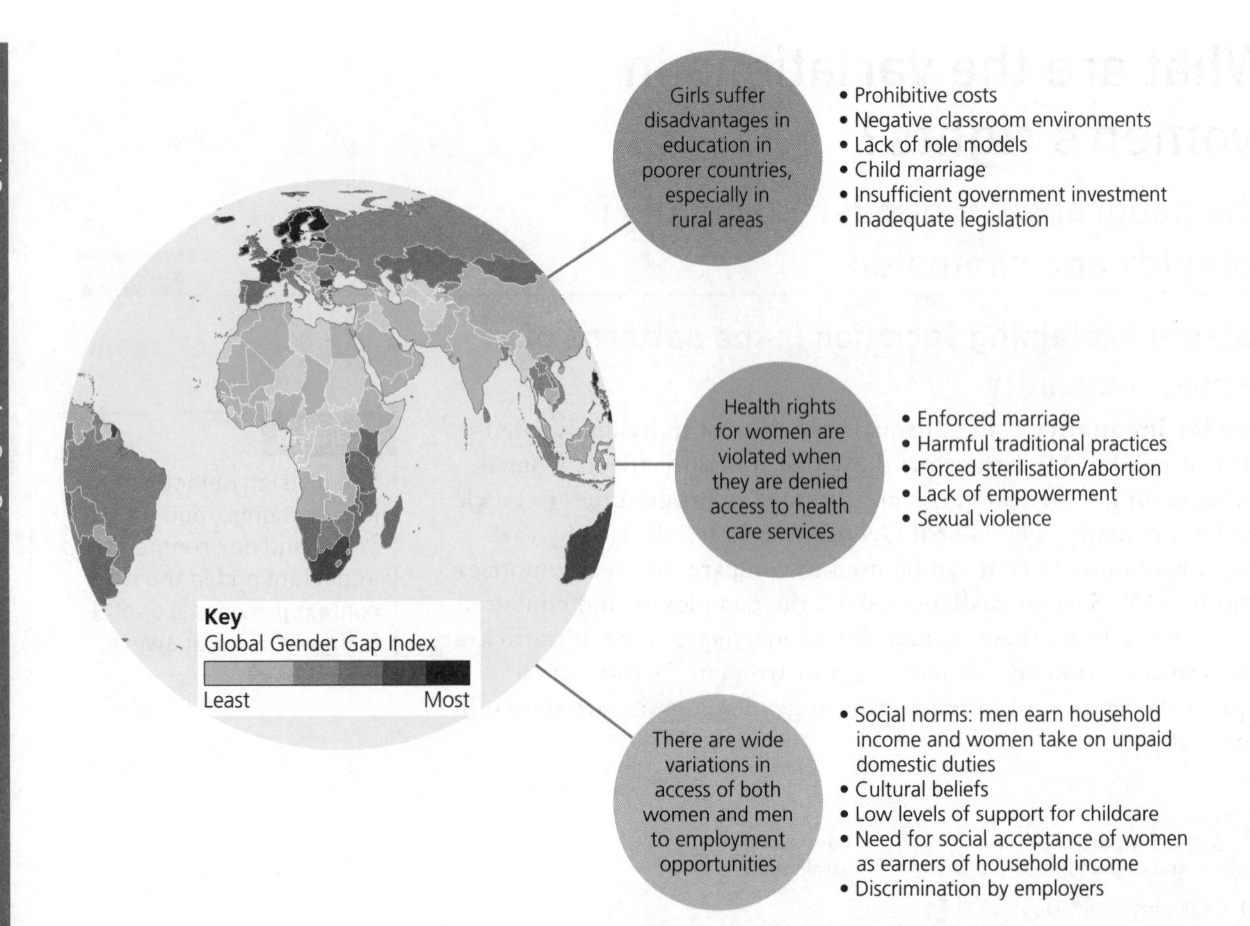

Figure 3.14 Factors explaining variation in the patterns of gender inequality

Case study of women's rights in a country

Online (see p. 3) you will find a case study illustrating India as an example of women's rights. The focus is on:

- the gender inequality issues of India
- the consequences of gender inequality for society
- evidence of changing norms and strategies to address gender inequality issues

Revision activity

Using Figure 3.14, (a) describe the global pattern of the gender map and (b) choose **three** country examples and annotate the map/diagram to explain the extent to which the gender gap is related to development.

What are the strategies for global governance of human rights?

Strategies for global governance of human rights include:

- laws and regulations: new laws are created and existing ones strengthened. **Treaties** and conventions are also established by large global organisations such as the UN, the EU and NATO
- the role of the UN peacekeeping operations
- humanitarian intervention and relief assistance
- attempts to change and modernise **norms**
- the influence of MNCs in terms of their Corporate Social Responsibility (CSR)
- the work of NGOs (non-government organisations) and private organisations

A **treaty** is a formally concluded and ratified agreement between states.

Norms are moral principles, customs and ways of living that are universally accepted as standard behaviour.

Exam tip

Scale provides an important structure to a discussion of strategies used to tackle any issue. Responses often exist at the global to local scale.

Human rights violations can be a cause and consequence of conflict

REVISED

How the violation of human rights can be a cause of conflict

Violations of human rights leading to conflict can be complex and several factors can be a cause of conflict or a contributing element, e.g. denial of basic human needs of food, housing and work, discrimination, oppressive governments, genocide and torture.

Revision activity

Select an example of a case where human rights violations have been the *cause* of conflict and a case where they have been a *consequence* of conflict and make a brief revision summary.

How the violation of human rights can be a consequence of conflict

Further human rights violations resulting from conflict may include: mortality, damage to homes, damage to infrastructure, impact on food and water supplies, displacement of people, exploitation of women and 'ethnic cleansing'.

Geopolitical intervention can include, for example, sanctions or military intervention.

Typical mistake

Rarely are issues of human rights violations straightforward. It is important to reflect the complexity of any situation.

The role of flows of people, money, ideas and technology in geopolitical intervention

Intervention involves flows of people and money to affected areas, e.g. personnel needed in peacekeeping missions. Significant amounts of money are also needed to fund operations, amounting to millions of dollars; the funds are supplied by member state contributions, the USA being the greatest contributor. An example of these costs is the stabilising mission in Haiti which has more than 4,500 personnel and access to over US$500 million donated by 51 countries.

There needs to be an efficient flow of ideas and information at the planning stage for intervention to be effective. The UNHRC employs individuals and working groups to promote ideas and values. Many NGOs, such as Amnesty International, publish information to increase awareness of human rights abuses.

Technology such as social media help with the flow of ideas. Remote sensing and satellite imagery are also used for surveillance and observation in dangerous areas.

Global governance of human rights involves cooperation at scales from global to local

REVISED

How human rights are promoted and protected by institutions, treaties, laws and norms

Institutions include the UN at a supranational scale, ASEAN at a regional scale, and national governments and NGOs working at the local scale.

- **UN:** an intergovernmental organisation with 193 member states working under the UN Charter. Human rights are at the centre of the Charter. The Office of the High Commissioner for Human Rights has the lead responsibility within the organisation. Council and treaty bodies work with the legal backing of the International Bill of Rights.

The Security Council deals with serious violations of human rights in conflict areas.

- **NGOs:** many NGOs are involved in the protection of human rights. They often work at the local scale to monitor situations and provide education and training programmes to support people.
- **Treaties, laws and norms:** a **treaty** is a written international agreement between two or more states of organisations. By signing or ratifying a treaty it becomes bound by international law. **International law** defines the responsibilities of states. Treaties and laws are defined from **norms** — long-established practices in many countries set out in the UN Charter, they are reinforced by treaties.

International law: body of law that governs international relations between states and nations.

Now test yourself

TESTED

36 What is the role of technology in geopolitical intervention?
37 Define 'norms'.

Answers on p. 220

Case study of strategies for global governance of human rights in one area of conflict

Revision activity

Make a revision summary of a case study covered in class for this section of the course. Remember to focus on:

- contributions and interactions of different organisations at a range of scales from global to local
- consequences of global governance of human rights for local communities

To what extent has intervention in human rights contributed to development?

Human rights are an essential part of achieving development. These strong links are shown in the UN's Millennium Development Goals (MDGs) and Sustainable Development Goals (SDGs).

Global governance of human rights has consequences for citizens and places

REVISED

How the global governance of human rights issues has consequences for citizens and places

Human rights can be regulated through global governance, states can report human rights violations committed within their countries to an external body such as the UN. This has can have consequences for civilians and places both in the short and longer term which are both positive and negative in nature.

Examples include:

- Military protection; provision of shelter, medicines, food and water (**short-term positive**)
- Infrastructure redevelopment; democratic elections; support in education and training; improved future food security (**long-term positive**).
- Population displacement; civilian casualties; destruction of property and infrastructure (**short-term negative**).
- Aid dependency; prolonged military and political conflict; significant impacts on population structure; long lasting negative place perception (**long-term negative**).

Exam tip

Make sure that you can give a cause and effect explanation of the consequences of the global governance of human rights issues for citizens and places and support your answer with named examples.

Now test yourself

TESTED

38 How can global governance affect human rights in the short term?
39 How can military intervention have a positive impact on human rights?

Answers on p. 220

Case study of the impact of global governance of human rights in an LIDC

Revision activity

Make a revision summary of a case study covered in class for this section of the course. Remember to focus on:

- the human rights issue/issues
- the global governance strategy/strategies used
- opportunities for stability, growth and development
- challenges of inequality and injustice

Exam tip

Remember to make case study information evaluative rather than purely a description or list of facts.

Summary

- Key terms to understand include **norms**, **intervention** and **geopolitics**.
- **Human rights** has a specific meaning and there are global variations in human rights norms.
- Several factors influence patterns of human rights violations.
- The geography of gender inequality is complex and contested.
- Human rights violations can cause conflict and can be a consequence of conflict.
- Global governance of human rights exists at a range of scales and includes the participation of a number of organisations.
- The global governance of human rights has short- and long-term impacts for people and places.

Exam practice

14 'Global governance of human rights presents more challenges than opportunities.' Discuss. [16]
15 'Economic factors are the most important influence on patterns of human rights violations.' Discuss. [16]

Answers and quick quiz 3C online

ONLINE

Option D Power and borders

What is meant by sovereignty and territorial integrity?

The world political map of sovereign nation-states is dynamic

REVISED

A world political map shows sovereign nation-states: independent self-governing countries. The dynamic nature of the map is shown by a range of examples, e.g. Eritrea gained independence from Ethiopia in 1993; in 1991, 15 new countries formed by **secession** from the USSR; and the area of Yugoslavia has seven countries today — the Balkan states. Any border changes affect sovereignty over populations and resources, the economy and the social geography of an area.

The world political map also shows disputed international borders.

Definitions

- **State:** an area of land of an independent country with well-defined boundaries, within which there is a body of people under a single government. States have several key features:
  - defined territory
  - sovereignty
  - government recognised by other states
  - capacity to engage in formal relations with other states
  - independence: self-governing
  - a permanent population with a right to self-determination

  State power depends on a wide range of economic, social, political and physical factors. The degree of resilience of a state is measured by the Fund for Peace (FFP) Fragile States Index. This measure includes a number of social, political, economic and political indices, e.g. refugees per capita and political prisoners.
- **Nation:** a large group of people with strong bonds of identity. A nation is different to a state in that it may be confined to one country, or its people may live in an area across adjoining countries; some may also be spread globally in the diaspora. States often contain several national groups. Nations do not have sovereignty.
- **Nation-state:** a nation that has an independent state of its own. In modern times, nearly all states refer to themselves as nation-states.
- **Sovereignty:** the absolute authority which independent states exercise in the government of the land and people in their territories.
- **Internal sovereignty:** a state's exclusive authority.
- **External sovereignty:** the need for mutual recognition among other sovereign states.
- **Territorial integrity:** states exercise their sovereignty within a specific territory, the boundaries of which have been established by international law. The preservation of territorial integrity and sovereignty is essential in achieving global stability and international security. A key organisation in maintaining this is the United Nations (UN).

Secession is transfer of part of a state's area and population to another state.

Exam tip

You should have a good working knowledge of the world political map.

Now test yourself

TESTED

40 What is the difference between a nation and a state?
41 What is the difference between internal and external sovereignty?
42 What is the difference between sovereignty and territorial integrity?

Answers on p. 221

Understanding of terms

- **Norms:** these are derived from moral principles and behaviours which have developed over time, throughout the world. They are part of international law and are upheld by state governments and citizens. They are based on the principles of the Charter of the United Nations: examples include Article 2.1 that allows all member states of the UN to have equal rights to determine their own form of government, chosen without outside influence. Governments are expected to put into place domestic laws in line with the UN Charter. There are an increasing number of 'regional' norms ratified by member states such as the EU.
- **Intervention:** this is sometimes required to resolve conflicts or humanitarian crises such as a serious violation of human rights. Examples include economic sanctions, military intervention and humanitarian assistance. Intervention is controversial as it is deemed to undermine the principle of sovereignty.
- **Geopolitics:** this relates to the global balance of political power. This power is very uneven: there are powerful ACs (e.g. the USA), increasingly influential EDCs (e.g. India), and LIDCs which remain peripheral. Multinational organisations such as the UN, the EU and OPEC also exert geopolitical power. TNCs are becoming influential as globalisation spreads.
- **Governance:** this is the application of laws, regulations, ethical standards and norms. **Global governance** is the way global affairs are managed. It usually works as a process of cooperative leadership whereby agreements which affect national governments are agreed. Recent foci of such agreements have been the environment, trade, poverty reduction, human rights, civil conflict, health issues and finance. Global governance is seen as the most effective way to achieve sustainable development in an interdependent world. However, critics say that it undermines the sovereignty of nations and marginalises poorer countries.

Exam tip

Intervention is a broad term; be specific if you use this term in exam answers.

Now test yourself

TESTED

43 What is a norm?
44 What is global governance?
45 What are the main criticisms of global governance?

Answers on p. 221

What are the contemporary challenges to sovereign state authority?

A multitude of factors pose challenges to sovereignty and territorial integrity

REVISED

Factors influencing the erosion of sovereignty and loss of territorial integrity

Current political boundaries

Current political boundaries are based on the Westphalian model based on the sovereign equality of all states. The principles are reinforced in the UN Charter but, increasingly, control of borders has been contested. Situations include:

- Russia's annexation of Crimea and support for separatists in Ukraine
- contested islands in the South and East China Seas
- claims of secession by Basque and Catalan national groups in Spain
- sectarian conflicts in the Middle East and North Africa
- transnational movement of terrorist and extremists, e.g. Syria
- contested maritime boundaries
- the legacy of colonialism where arbitrary political boundaries in Africa split ethnic groups, e.g. Mali

Transnational corporations (TNCs)

TNCs characteristically operate in more than one country. Growing in number, they have increasing influence in the global economy and the integration process of globalisation. The economic development that TNCs bring to LIDCs through much-needed investment also brings disadvantages. Sometimes they present a challenge to state sovereignty and government control. Further criticisms centre on a disrespect for human rights by some TNCs, e.g. exploitation of workers and the fact that business decisions that impact many people have been taken outside of the host country.

Supranational institutions such as large regional trading blocs

Supranational institutions represent a level of governance beyond that of an individual state. Examples include the UN and trading blocs such as the EU. In addition to their own independence, member states are also bound by the requirements and treaties of the institution to which they belong. Criticism of such membership centres on the 'surrender' of some aspects of a country's sovereignty. Those in favour of Brexit supported this critical view. Table 3.9 summarises the features of two supranational institutions, the EU and the UN.

Supranational: having influence and power that transcends national boundaries or governments.

Table 3.9 **How sovereignty is affected by membership of the EU and UN**

EU	UN
• A large trading bloc and a political union with a parliament and monetary 'Eurozone' • Currently 28 sovereign member states • Benefits of integration and access to a large trade area • Costs as there are some challenges to a country's individual autonomy such as having to implement laws even if they did not vote for them • Additional financial restrictions for those in the Eurozone, e.g. austerity measures for Greece's debt crisis • Other limitations include fishing quotas through the Common Fisheries Policy and regulations on farming and the countryside through Common Agricultural Policy • Migration laws are also contentious — the Schengen Agreement abolished many of the EU's internal borders, enabling passport-free movement across most of the bloc	• 193 sovereign states are members of the General Assembly • Sovereignty and integrity are upheld in the UN Charter (Articles 2.1 and 2.4) • The state has direct responsibility for its bounded territory • However, the UN with the backing of the Security Council can sanction intervention • Intervention applies in instances where there are considerable crimes against humanity, e.g. genocide and ethnic cleansing

Political dominance of ethnic groups

- Ethnic groups are spread across political borders.
- A sovereign state may include several ethnic groups, e.g. Sudan has an estimated 60 different ethnic groups.
- A single ethnic group may be portioned by modern state borders, e.g. the Tuareg homeland extends across five states in North and West Africa.
- Ethnic groups challenge a state when they have strong identities and demand independence in a new state, e.g. the Tuareg in Mali.
- Sovereignty is also challenged by internal conflicts between ethnic groups.

Now test yourself

TESTED

46 State **two** challenges to state sovereignty from TNCs.
47 How do ethnic groups challenge sovereignty?

Answers on p. 221

Case study of one country in which sovereignty has been challenged

Online (see p. 3) you will find a case study on Ukraine. This shows how you could make summary revision notes of a case study for this part of the course of a country in which sovereignty has been challenged.

What is the role of global governance in conflict?

Global governance provides a framework to regulate the challenge of conflict

REVISED

How challenges to sovereignty and territorial integrity can be a cause of conflict

Challenges to sovereignty and/or territorial integrity arise in a variety of situations, e.g. competition over scarce resources, religious persecution, failure of governments to protect basic human rights.

The Global Peace Index (a composite index including level of safety and degree of militarisation) represents conflict situations. A high score means a less peaceful nation and a low score is more peaceful.

Revision activity

Using an up-to-date version of the interactive Global Peace Index map **(www.visionofhumanity.org)**:

(a) describe the overall pattern
(b) in general terms list potential reasons for the location of more and less peaceful areas
(c) select an example of **one** country at each extreme of the continuum and account for its position

The role of institutions, treaties, laws and norms

Examples of institutions involved in the intervention process in order to regulate conflict and maintain the global system of sovereign nation-states include the UN, NATO (the North Atlantic Treaty Organization), the EU, ASEAN (the Association of Southeast Asian Nations) and many NGOs (non-government organisations).

Revision activity

Institutions with a role in the regulation of conflict operate at different spatial scales. Table 3.10 shows a summary of the role of the UN, an international institution. Select one regional organisation such as the EU and one NGO such as Amnesty International and complete Table 3.10 with a summary of their roles.

Table 3.10 The role of institutions in regulating conflict and in maintaining the global system of sovereign nation-state

The UN (operational at the international scale)	(A regional institution)	(An NGO)
1 Founded in 1945, HQ in New York, historic home Geneva, the UN is an important global partnership 2 The UN aims to: – achieve world peace and security – develop good relations between nations – foster cooperation among nations 3 The UN Charter gives the Security Council responsibility to maintain international peace and security. It does this through, for example: treaty-making, peacekeeping, delivering humanitarian aid and upholding international law 4 Preventative diplomacy and mediation and also monitoring and observation are crucial elements of the work of the UN		

Treaties, laws and norms

A **treaty** is a written international agreement between two or more states of organisations. By signing or ratifying a treaty it becomes bound by international law.

International law defines the responsibilities of states and regulates conflict over **global commons**.

Treaties and laws are defined from **norms** — long-established practices in many countries set out in the UN Charter, they are reinforced by treaties.

International law is the body of law that governs international relations between states and nations.

The **global commons** are the Earth's shared natural resources, e.g. oceans and atmosphere.

The role of flows of people, money, ideas and technology in geopolitical intervention

The work of the UN and any other organisation intervening in areas of conflict requires the movement of people and resources. These resources can reach high levels — thousands of staff and billions of dollars.

Effective intervention also relies on a constructive exchange of ideas from gathering of intelligence to regional and global meetings and strategic planning.

Technology plays an increasingly key role in peacekeeping for planning and surveillance. Examples include satellite imagery, use of drones, mobile telephony, social network services and the use of international databases.

Exam tip

Always remember to present a balanced and fair view in discussions of geopolitical conflict and intervention.

Now test yourself TESTED

48 What are the 'global commons'?
49 What is the role of monetary flows in geopolitical intervention?

Answers on p. 221

Exam tip

Scale is a key concept in geography and can provide a useful structure to exam answers.

Global governance involves cooperation between organisations at scales from global to local, often in partnership

REVISED

Case study of strategies for global governance in one area of conflict

Revision activity

Make your own summary revision notes of an example studied in class for this part of the course. Remember to focus on the following:

- interventions and interactions of organisations at a range of scales
- consequences of global governance of the conflict for local communities

How effective is global governance of sovereignty and territorial integrity?

Global governance of sovereignty and territorial integrity has consequences for citizens and places

REVISED

Global governance aims to stabilise governments when sovereignty and territorial integrity are threatened and to ensure that citizens' rights and socio-economic wellbeing is protected. Intervention has consequences for both citizens and places. These can be short or long term and include positives such as aid, shelter and food security. Negative impacts may include civilian casualties, population displacement and an escalation of violence.

Revision activity

Produce a series of revision tables or diagrams to list the consequences for citizens and places of the global governance of:

a) sovereignty issues
b) territorial integrity

Categorise the consequences as positive or negative and short or long term.

Remember to be able to identify economic/social/environmental and political consequences.

Now test yourself

TESTED

50 State:
(a) **two** long-term benefits of global governance where sovereignty has been threatened
(b) **two** short-term benefits of global governance where territorial integrity has been threatened

Answers on p. 221

Case study of the impact of global governance of sovereignty or territorial integrity in one LIDC

Revision activity

Make your own summary revision notes of an example studied in class for this part of the course. Remember to focus on the following:

- the sovereignty or territorial integrity issue/issues
- the global governance strategy/strategies used
- opportunities for stability, growth and development
- challenges of inequality and injustices

Summary

- The world political map of sovereign nation states is dynamic.
- Understanding of key terms includes definitions of **state**, **nation**, **sovereignty** and **territorial integrity**.
- A range of factors challenge sovereignty and territorial integrity; they can be categorised into economic, political, social and environmental.
- Conflict arises when sovereignty and territorial integrity are challenged.
- A range of institutions, treaties, laws and norms regulate the challenge of conflict.
- Global governance operates at a range of scales and with a variety of strategies which can be illustrated through case study evidence.
- Global governance of sovereignty and territorial integrity issues has consequences for people and places.

Exam practice

16 'TNCs pose the greatest challenge to sovereign state authority.' Discuss. [16]

17 'Global governance of territorial integrity issues brings considerable long-term benefits.' Discuss. [16]

Answers and quick quiz 3D online

ONLINE

4 Geographical debates options

Option A Climate change

How and why has climate changed in the geological past?

The Earth's climate has undergone huge change in the past 100 million years, particularly in the Quaternary (last 2.5 million years). Today, concern centres on a warming climate with shrinking ice sheets and glaciers and rising temperatures.

The Earth's climate is dynamic

REVISED

Methods used to reconstruct past climate

Table 4.1 summarises the main methods used to reconstruct past climate.

Table 4.1 Methods used to reconstruct the Earth's past climate

Sea-floor sediments	The fossil shells of tiny sea creatures called foraminifera, which accumulate in sea-floor sediments, can be used to reconstruct past climates. The chemical composition of foraminifera shells indicates the ocean temperatures in which they formed
Ice cores	Ice cores from the polar regions contain tiny bubbles of air — records of the gaseous composition of the atmosphere in the past. Scientists can measure the relative frequency of hydrogen and oxygen atoms with stable isotopes. The colder the climate, the lower the frequency of these isotopes
Lake sediments	Past climates can be reconstructed from pollen grains, spores, diatoms and varves in lake sediments. Pollen analysis identifies past vegetation types and from this infers palaeoclimatic conditions. Pollen diagrams show the number of identified pollen types in the different sediment layers. Diatoms are single-celled algae found in lakes, with cell walls made of silica. They record evidence of past climates in their shells (see foraminifera above). Varves are tiny layers of lake sediment comprising alternating light and dark bands. The light bands, formed from coarser sediments, indicate high-energy meltwater run-off in spring and summer. The darker bands, made up of fine sediment, show deposition during the winter months
Tree rings	Dendrochronology is the dating of past events such as climate change through study of tree ring (annule) growth. Annules vary in width each year depending on temperature conditions and moisture availability
Fossils	Plants and animals require specific environmental conditions to thrive. Some, such as coral reefs, are highly sensitive to temperature, water depth and sunlight. Where they exist in the fossil record they can be used as proxies for climate. Animals are more adaptable. However, some herbivorous dinosaur species only survived in sub-tropical habitats

Past climate to reveal periods of greenhouse and icehouse Earth

Long-term changes

About 100 million years ago (mid-Cretaceous) average global temperatures were 6–8°C higher than today and CO_2 levels were five times higher. A further, shorter spike in temperatures occurred 55 million years ago in the Palaeocene–Eocene thermal maximum, when average global temperatures were 23°C. There was a change to colder conditions 35 million years ago.

Glaciation of Antarctica

The continent of Antarctica is covered with thick ice so that only the highest mountains appear above the ice — but 40 million years ago the continent had sub-tropical conditions. The change occurred because:

- CO_2 levels dropped 35 million years ago
- continental drift led to isolation of the continent in the South Pole
- ocean currents isolated the continent from warm water

Quaternary glaciation

The Quaternary period is characterised by cyclic changes in climate from cold glacials to warmer interglacials. The most recent glacial was around 20,000 years ago when about one-third of the continental surface was covered with ice. As the climate warmed the ice remained only in mountains until *c.* 13,000 BP (before present).

Present interglacial, the Holocene

An interglacial period separates glacials, which have dominated 90% of the Quaternary. Except for Antarctica and Greenland, glaciers and ice sheets have shrunk. Ice fields and valley glaciers remain only in the highest mountains, e.g. the Himalayas, and high latitudes, e.g. Alaska. There have been fluctuations within this warming period. Scientists believe that the climate change of the last 200 years has been driven by human activity, entering us into a new period — the **Anthropocene**.

Typical mistake

Because of the complexity of the global climatic system, there will always be exceptions to overall trends.

How natural forcing has driven climate change in the geological past

Table 4.2 summarises the driving forces of climate change in the geological past.

Table 4.2 The driving forces of climate change in the geological past

Driving force	Description
Plate tectonics and volcanic eruptions	Because of plate tectonics and sea-floor spreading, the distribution of the continents has changed. The scale of the changes impacts climate — large continents across high latitudes increase global albedo and lead to global cooling Volcanic eruptions release ash and sulphur dioxide into the atmosphere. In the short term this has a cooling effect on climate (solar radiation is reflected back into space)

Driving force	Description
Milankovitch cycles	Milankovitch argued that climatic change is related to astronomical events such as changes in the Earth's axis and orbit, which in turn affect the amount of solar radiation reaching the Earth's surface. The cycles operate from 10 to 100,000 years
Solar output	The sun's solar output varies over time; sunspot activity shows a positive correlation with solar energy output. However, isolating the impact of variations in solar input from other influences, e.g. human activity, is difficult
Natural atmospheric greenhouse gases	There is a close relationship between atmospheric CO_2 and global temperature. Low levels of CO_2 reduce the Earth's natural greenhouse effect and lead to cooler climatic conditions. 50 million years ago there were large concentrations of CO_2 in the atmosphere (1,000 ppm) and global temperatures were 10°C higher Theories of how this CO_2 has been removed relate to plate tectonics — the mountains created increased rainfall charged with CO_2, and in this way CO_2 was removed from the atmosphere and transferred to ocean storage

Now test yourself

TESTED

1 What is a glacial and an interglacial period?
2 What is the Anthropocene?
3 State **three** natural forcing causes of climate change.

Answers on p. 221

How and why has the era of industrialisation affected global climate?

Humans have influenced the climate system, leading to the Anthropocene

REVISED

Evidence the world has warmed since the late nineteenth century

Increases in temperatures

- Rising temperatures have been a trend over the past 135 years.
- Anomalies occurred between 1880 and 2014.
- There has been a steep rise in temperatures in the twenty-first century — land temperatures 1°C above average and ocean temperatures 0.57°C above average.
- 2014 was the 38th consecutive year that annual global temperature rise was above average.

Shrinking of valley glaciers and ice sheets

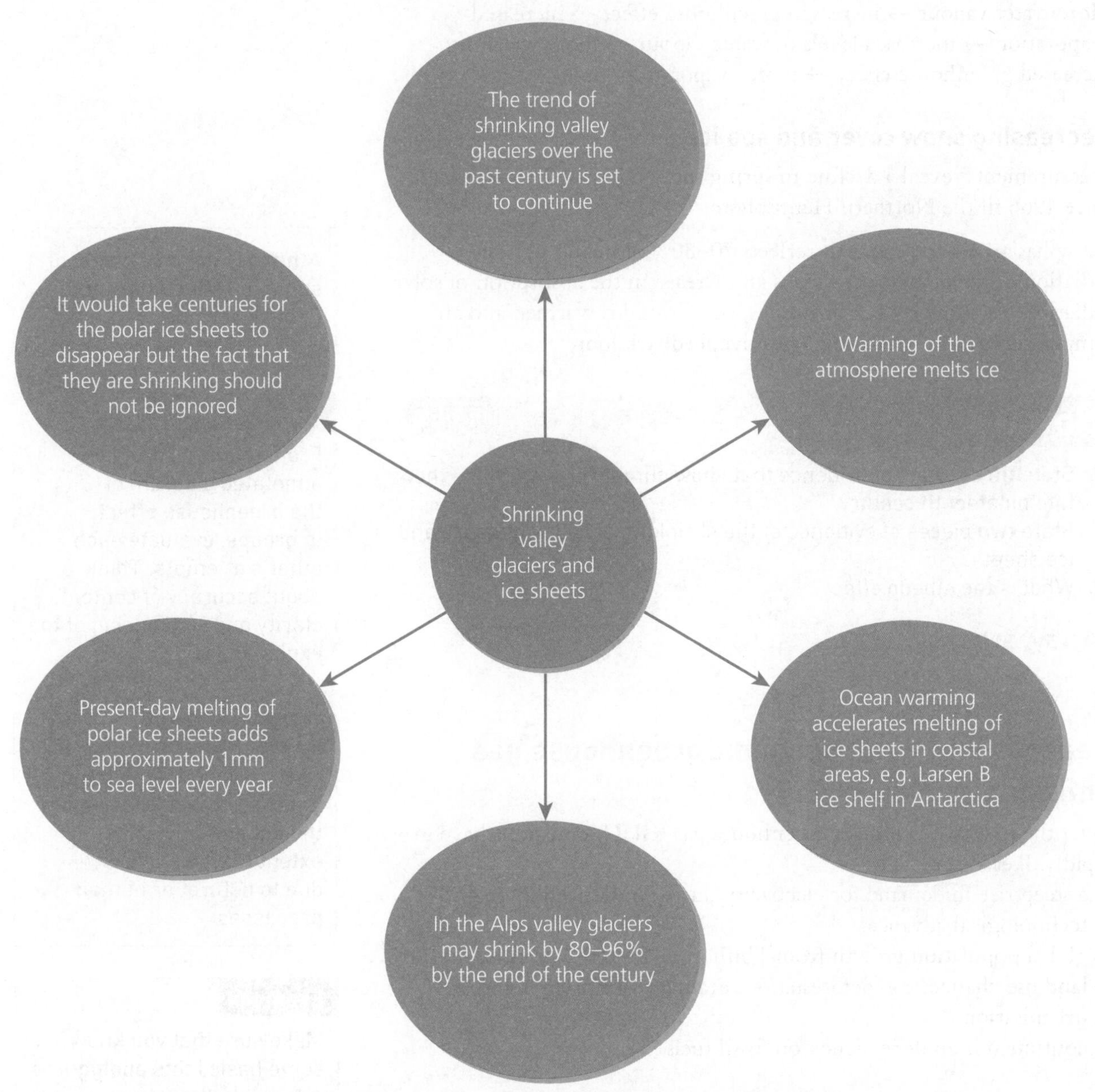

Figure 4.1 Shrinking of valley glaciers and ice sheets as evidence for climate change

Rising sea level

Since 1900 the average sea level rise has been 1.0–2.5 mm/year. Recent satellite data suggest a more rapid rate of 3 mm/year. Thermal expansion of the oceans and melting of land-based ice sheets and glaciers account for the rise.

Increasing atmospheric water vapour

Water vapour traps energy radiated from the Earth's surface and creates a **greenhouse effect**. Water vapour is affected by temperature (warm air holds more) and rates of evaporation (higher evaporation → more atmospheric water vapour). Increasing temperatures lead to more atmospheric water vapour.

Estimates suggest that for every 1°C increase in temperature, rising levels of water vapour will double the warming.

> The **greenhouse effect** occurs when carbon dioxide and other greenhouse gases act like a blanket, absorbing incoming radiation and preventing it from escaping back into the atmosphere. The net effect is the gradual heating of Earth's atmosphere and surface, a process known as global warming.

There is a dangerous **positive feedback** effect:

More water vapour → increased greenhouse effect → increased evaporation → increased levels of water vapour → more warming/increased greenhouse effect → more evaporation …

Decreasing snow cover and sea ice

Measurements reveal a decline in spring snow cover of 2% per decade since 1966 in the Northern Hemisphere.

Snow has an **albedo** effect (it reflects 70–80% of incoming solar radiation). Less snow cover means an increase in the absorption of solar radiation. As snow cover diminishes, the ground is warmed and air temperatures rise — forming a positive feedback loop.

Albedo is the proportion of sunlight reflected from the Earth's surface.

Now test yourself

TESTED

4 State **three** types of evidence that show climate change since the late nineteenth century.
5 State **two** pieces of evidence of the shrinking of valley glaciers and ice sheets.
6 What is the albedo effect?

Answers on p. 221

Revision activity

From memory, draw an annotated diagram of the greenhouse effect. In groups, evaluate each other's attempts. Think about accuracy of content, clarity and sequencing of the explanation.

Reasons why anthropogenic greenhouse gas emissions have increased

Over the past two centuries greenhouse gas (GHG) emissions have grown rapidly. Reasons include:

- a steep rise in demand for electricity due to industrialisation and technological advances
- global population growth from 1 billion in 1800 to 7.4 billion in 2015
- land use change, e.g. deforestation, drainage of wetlands and urbanisation
- continued high dependency on fossil fuels

Typical mistake

There is no doubt that the Earth's climate is warming; the debate is over the extent to which this is due to natural or human processes.

Exam tip

Make sure that you know some basic facts and figures relating to the reasons why GHG emissions have increased.

The balance of anthropogenic emissions around the world

Figure 4.2 shows a range of facts explaining how the balance of anthropogenic emissions around the world has changed.

Between 1850 and 1960 most GHG emissions came from the industrialised economies of North America and Europe → Since the 1960s emissions from Asia, particularly China, have increased significantly. Those of North America and Europe have stabilised and in some countries, e.g. Germany and the UK, they have actually fallen → Global emissions are now uneven but heavily concentrated with the top ten countries accounting for almost 80% of emissions. The USA, Australia, Germany and the UK still account for the highest CO_2 emissions → When emission of CH_4 from land-use change is added as a GHG, Brazil and Indonesia enter the ranking in third and fourth place after China and the USA

Figure 4.2 The changing balance of anthropogenic emissions around the world

How additional greenhouse gases being added to the atmosphere will enhance the natural greenhouse effect

GHGs such as water vapour, CO_2 and CH_4 occur naturally in the atmosphere. They allow short-wave radiation from the sun through but absorb the Earth's long-wave radiation, making the Earth's surface warmer. Figure 4.3 shows the greenhouse effect.

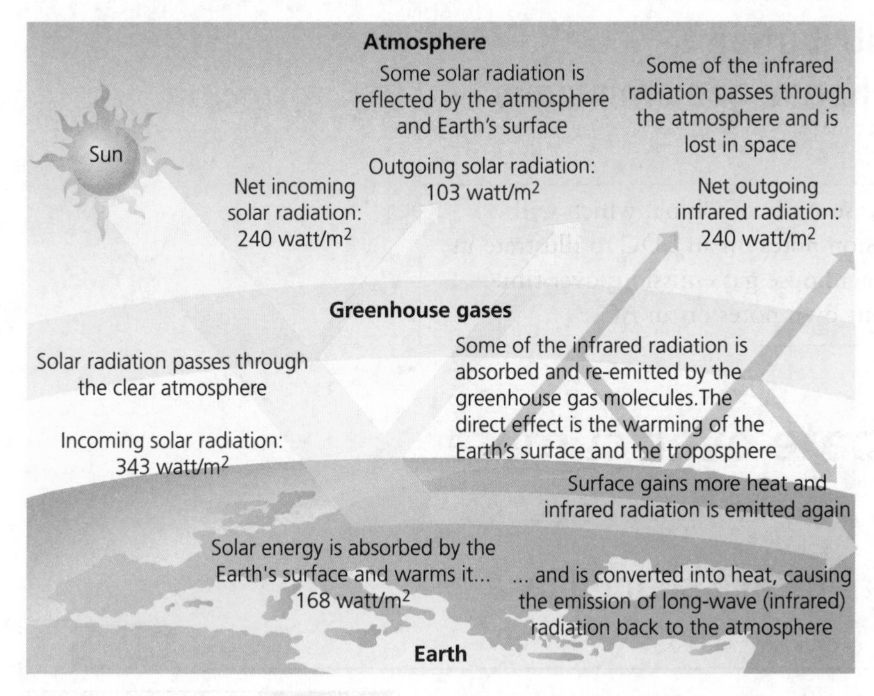

Figure 4.3 The greenhouse effect

Since the early nineteenth century, the volume of GHGs has increased, particularly CO_2, by one-third. The **enhanced greenhouse effect** is the impact on the climate from additional heat retained as a result of increased amounts of carbon dioxide and other GHGs that humans have released into the Earth's atmosphere since the Industrial Revolution.

How humans influence the global mean energy balance

- The Earth's atmosphere is a closed system; the system is stable when inputs of solar radiation equal outputs of terrestrial radiation. About 45% of incoming solar radiation reaches and warms the Earth's surface (one-third is reflected from clouds, one-fifth absorbed by water droplets, ozone, CO_2 and dust in the atmosphere, and a small amount is reflected from surfaces such as snow).
- Of the outgoing radiation, two-thirds are lost in space and the rest is absorbed by GHGs.
- GHGs play a crucial role in stabilising global temperatures.
- In the last 200 years human activity has upset the global energy balance: increasing amounts of GHGs absorb more terrestrial radiation; global temperatures have risen; rising temperatures increase evaporation and the amount of water vapour in the atmosphere; rising temperatures also melt snow and ice, reducing the reflection of incoming solar radiation and adding heat energy to the Earth.

Now test yourself

TESTED

7 What is the enhanced greenhouse effect?

Answers on p. 221

Case studies of one AC and one EDC to illustrate their contribution to anthropogenic greenhouse gas emissions over time

Online (see p. 3) you will find a case study of China, which will show how you could present revision notes on an EDC to illustrate its contribution to anthropogenic greenhouse gas emissions over time. Use this as a template to make your own notes on an AC.

Why is there debate over climate change?

Debates of climate change are shaped by a variety of agendas

REVISED

How humans have played a part in shaping the climate change debate

Historical background to the global warming debate

1842 Greenhouse effect discovered → **1862** suggestion that CO_2 and water vapour trapped heat escaping from the Earth's surface → **1896** suggestion that CO_2 absorbs long-wave radiation and that doubling CO_2 would increase temperatures by 5°C or 6°C → **1938** global warming linked to fossil fuel emissions → **1957** scientific progress ruled out the ability of oceans to absorb excess CO_2, meaning that CO_2 remains in the atmosphere for much longer than originally thought → **1958** databases showed proof that CO_2 concentrations in the atmosphere were increasing year by year (the Keeling curve) → **1970s** much debate when satellite imaging and computer modelling advanced the argument in favour of global warming → **1988** the Intergovernmental Panel on Climate Change (IPCC) was set up.

Debate has continued to the present day, with the majority of climate scientists agreeing that global warming is taking place. Such is the complexity of the issue, some doubt still exists.

Exam tip

The historical background is useful in giving context to the climate change debate. Constructing a timeline is a useful means of revision.

The role of governments and international organisations

Revision activity

Complete Table 4.3 with summary notes on the role of international organisations and governments; as an example, the IPCC has been done for you. Add one other international organisation, e.g. the EU or UN, and details from governments/countries with contrasting views, e.g. the UK and India.

Table 4.3 The role of governments and international organisations

Organisation or government	Role (for an organisation) or viewpoint (for a country) and criticisms (for an organisation) or evaluation (for a government)
IPCC	**Role** ● Purpose is to provide policy makers with an objective opinion on the causes of climate change, the potential environmental impacts and the possible response options ● Collects and assesses the best available scientific, technical and socio-economic information on climate change **Criticisms** ● A lack of specific guidance on how countries should lower emissions ● The IPCC has allegedly diluted the impacts of climate change as a result of political pressure from industrialised nations such as China, Brazil and the USA ● The process of reporting has been called into question after errors in some reports
	Role **Criticisms**
	Viewpoint **Evaluation**
	Viewpoint **Evaluation**
	Viewpoint **Evaluation**

The role of the media and different interest groups

- The role of the media, through TV, magazines and newspapers, is an influential one in forming public opinion.
- Most people do not read academic scientific journals and reports.
- The political 'leaning' of a newspaper is an important consideration, e.g. right-leaning *Times* and left-leaning *Guardian*.
- TV outlets can also face criticism of bias in their reporting.
- The role and viewpoint of industries, such as energy and mining industries, adds a further element to the debate.

Revision activity

Use a suitable format such as a counterbalance (Figure 4.12, p. 153) or a comparison diagram (Figure 4.31, p. 186) to make revision notes on specific examples of the different viewpoints contributing to the debate on climate change.

In what ways can humans respond to climate change?

An effective human response relies on knowing what the future will hold

REVISED

An overview of climate modelling

Importance of the carbon cycle

An overview of the carbon cycle is shown in Figure 4.4. Within this there are four 'subsystems': the terrestrial or 'fast' carbon cycle, the oceanic carbon cycle, the atmospheric carbon cycle and the 'slow' carbon cycle. The main processes in the carbon cycle, shown on Figure 4.4, are respiration, photosynthesis, decay/decomposition, weathering and, since industrial times, combustion.

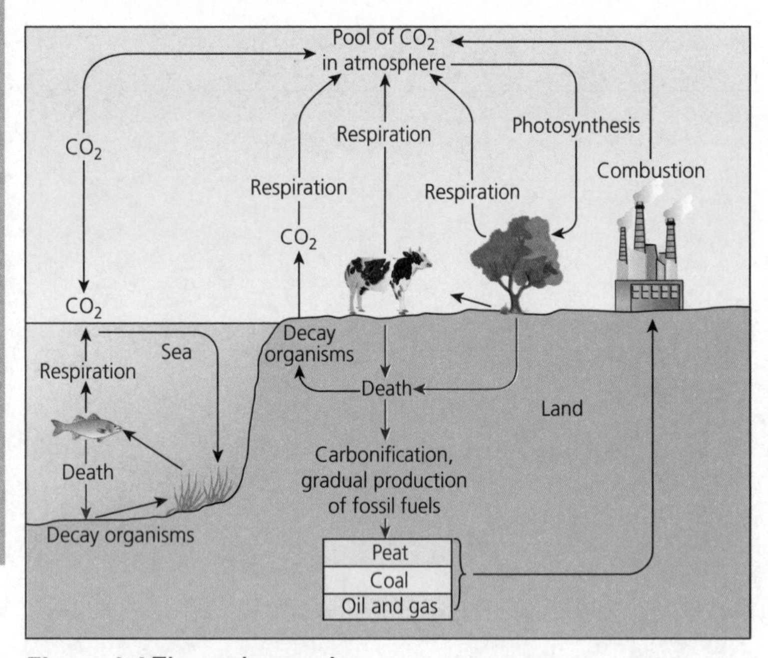

Figure 4.4 The carbon cycle

Exam tip

You should have knowledge of the carbon cycle and its basic functioning so you can discuss its importance. Carbon as a building block to life on Earth is covered on p. 58.

Carbon is important to life on Earth because:

- all living organisms depend on carbon
- green plants and **phytoplankton** are **primary producers** in ecosystems which convert sunlight and CO_2 to carbohydrates in the process of photosynthesis; this supports all consumer organisms
- decomposition and oxidation ensure the recycling of CO_2 and the replenishment of vital stores
- CO_2 is a key component of the atmosphere, absorbing long-wave radiation

Phytoplankton are tiny photosynthesising marine organisms living in the surface waters of oceans where there is most light.

Primary producers are green plants that convert sunlight, carbon dioxide and mineral nutrients into organic matter (chemical energy) by the process of photosynthesis.

Positive or **negative feedback**: feedback is a response that changes a system; it can be positive when it generates further change or negative when it restores equilibrium.

The influence of positive and negative feedback

Although CO_2 contributes less to the overall greenhouse affect than does water vapour, it is CO_2 that sets the temperature as it affects radiation in the atmosphere and causes an overall warming effect. The warming affects the amount of water vapour in the atmosphere. Climatic feedbacks seek to explain how systems react to that direct warming either as a **positive** or **negative feedback** (Figure 4.5).

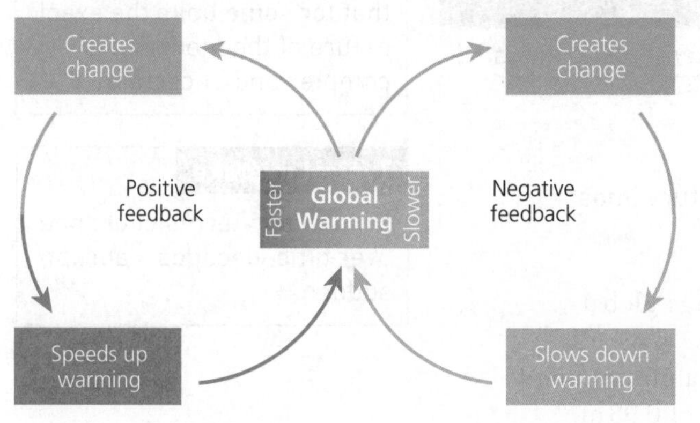

Figure 4.5 Climatic feedback

Climatic feedbacks can have a significant impact on the magnitude of potential future climate change. Examples are given in Table 4.4.

Table 4.4 The influence of positive and negative feedbacks

Change	Feedback effect
Water vapour feedback	An increase in CO_2 leads to an increase in the temperature of the atmosphere, warm air holds more water vapour, water vapour helps the Earth hold on to more heat energy from the sun = warmer climate. **Positive feedback**
Albedo feedback	Bright surfaces, increased reflection of sunlight, but with a warmer climate snow melts, less sunlight is reflected, more warming = warmer climate. **Positive feedback**
Clouds	Complex. Clouds reflect sunlight back to space leading to cooling, but clouds can also restrict heat radiated back leading to warming. The key is in the fact that high clouds retain heat for longer, while low cloud is thicker and reflects more sunlight. In a warming world we may get more high cloud. **Some positives and negatives**
Terrestrial carbon cycle	Some CO_2 in the atmosphere is absorbed by plants and the soil, some by oceans. Changes to land surfaces in the future will affect CO_2 release and absorption, e.g. soils may be warmer and release CO_2. **Some positives and negatives**
Very unsure: Feedback of methane hydrates: there is a large stock of methane in the deep ocean. In a warming ocean this **may** be released Permafrost areas: carbon locked in organic-rich soils in permafrost areas in high latitudes **may** be released in a warming climate	

Now test yourself

TESTED

8 Why does the influence of clouds on climate have both positive and negative feedbacks?

Answers on p. 221

> **Exam tip**
>
> Practise your application of the knowledge in Table 4.4 to different forms of data presentation, e.g. composite graphs showing changes over time or linking two factors.

> **Exam tip**
>
> The concept of feedback is very important in the study of climate change. Be clear whether the response is positive or negative feedback and remember that for some links the exact nature of the feedback is complex and uncertain.

> **Typical mistake**
>
> Note that in terms of change over time, decades = abrupt/sudden.

Future emission scenarios and the impact on global temperatures and sea levels

Forecasting the future is extremely difficult because of the complex nature of the Earth's atmospheric system and uncertainties about future GHG emissions. The IPCC therefore provides a range of forecasts. Table 4.5 shows the IPCC's four scenarios.

Table 4.5 IPCC forecasts on GHG emissions

Description
GHG emissions peak 2010–2020 and decline thereafter (most optimistic)
GHG emissions peak around 2040 and then decline
GHG emissions peak around 2080 and then decline
GHG emissions rise throughout the twenty-first century (most pessimistic)

- All IPCC projections show significant rise in average global temperatures during the twenty-first century.
- Current projections for global sea level change are a minimum of 0.28 m by the end of the century, and a maximum of 0.98 m.

The impacts of climate change are global and dynamic

REVISED

Implications of climate change currently being experienced for people and the environment

Current impacts on people include the following.

- Despite the fact that between 2000 and 2015 global malaria mortality rates fell by 60% because of better treatment and prevention, current climate change has stimulated transmission of vector-borne diseases and extended their geographic spread. The concern is that with warmer climates the disease could spread to parts of the world that are currently malaria-free, e.g. southern Europe.
- Human health can also be compromised by droughts and floods reducing food production. This already threatens food security and human health in countries such as Yemen, South Sudan, Nigeria and Somalia where physical and human causes combine to impact food supply. Somalia is currently experiencing the worst drought for 36 years.
- Extreme weather events such as floods, tropical cyclones and heatwaves have implications for people, the economy and society and the concern is that climate change will lead to more extreme weather. Countries across the development spectrum have been affected in recent times, e.g. hurricane Katrina in 2003 which caused economic damage and loss of life in New Orleans and a prolonged heatwave in Europe, causing an estimated 35,000 excess deaths.

> **Exam tip**
>
> Remember that the impacts of climate change will vary considerably across the world, with factors such as economic status, current climate and ecosystem fragility.

Future projections

The IPCC projects significant reductions in staple crops by 2030 (Figure 4.6).

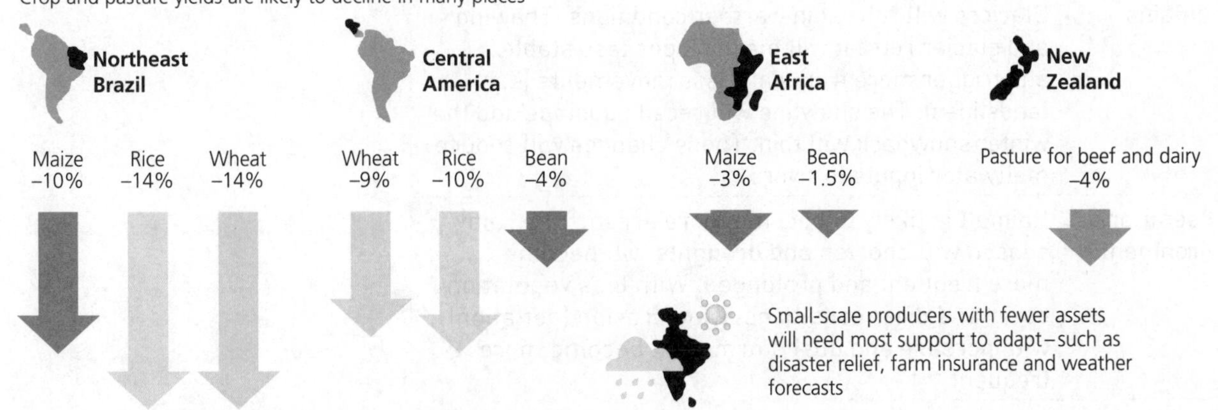

Figure 4.6 The future of farming and food

> **Exam tip**
>
> Reliable future predictions are extremely difficult given the complex nature of the climate change issue.

Current impacts on the environment

- Coral reefs are currently threatened by **coral bleaching**. In the last 30–40 years Indonesia has lost half of its reefs to bleaching and the Caribbean has lost 80%.
- Massive changes to marine ecosystems are currently taking place in the Arctic as a result of ocean warming and shrinking sea ice. This has impacted the delicate food chains: for example, there has been a two-thirds decline in the polar bear population. Around the UK, average sea surface temperatures have risen by 1.6°C, affecting the fishing industry as cold-water fish move northwards.
- Terrestrial ecosystems are also being impacted. As permafrost thaws in the tundra there are impacts on biodiversity, migration patterns, feeding and natural habitats.
- The IPCC has also recorded that in the past 30 years spring has occurred earlier by 2.3–5.2 days per decade on average, creating a loss of synchronisation between species.

> **Coral bleaching** occurs when the algae that give the coral its colour are lost because of an increase in water temperature.
>
> **Land degradation** is a process leading to a significant reduction in the production capacity of land.
>
> **Desertification** is a reduction in agricultural productivity due to overexploitation of resources and natural processes such as drought.

The vulnerability of people and the environment to the impacts of climate change

Vulnerability of people is affected by where they live and their ability to cope. Rural communities in the developing world, where most people rely on subsistence farming, are extremely vulnerable, especially where they already farm marginal land, e.g. the severe **land degradation** and **desertification** in the Sahel, northern Africa.

Other vulnerable farming areas include the Prairies in North America and the Pampas in Argentina. Declining yields in these areas could lead to global shortages of wheat and maize. Also, rural regions of northern India that are fed by glacial meltwater, and low-lying coastal regions in the tropics and sub-tropics, are vulnerable to flooding, e.g. Bangladesh.

Vulnerable environments are already fragile ecosystems, such as the tundra, mountains, deserts and the Amazon rainforest. These are summarised in Table 4.6.

Table 4.6 Environments most vulnerable to climate change

Environment	Change
Tundra	Rising temperatures melt the permafrost, disrupting vegetation, creating extensive thaw lakes and wetlands, and initiating mass movement. Forests will invade the southern margins of the tundra
Mountains	Glaciers will retreat in warmer conditions. Thawing and glacier retreat will make slopes less stable and trigger more frequent mass movements (e.g. landslides). The snowline will recede upslope and the winter snowpack will thin. These changes will reduce meltwater inputs to rivers
Hot semi-arid environments	Rainfall is likely to become more erratic, the rainy season will shorten and droughts will become more frequent and prolonged. With less vegetation cover and drier conditions, wind erosion (deflation) will increase and dust storms will become more frequent
Rainforest	Computer models predict that the Amazon rainforest will become warmer and drier by the mid twenty-first century. As deforestation increases (due partly to climate change and partly to clearance for agriculture, ranching and logging), the water cycle will weaken, creating positive feedback and accelerating forest loss — 30–60% of the Amazon rainforest could become dry savanna grassland by the end of the century
Coasts	Higher sea levels and more powerful storms will increase rates of erosion on both upland and lowland coasts. Shorelines will retreat inland. Coastal environments at particular risk include dunes, salt marshes and mudflats

Now test yourself

TESTED

9. Why is the Arctic particularly affected by climate change?
10. Why will climate change have potentially the greatest human impacts in Africa?

Answers on p. 221

Mitigation and adaptation are complementary strategies for reducing and managing the risks of climate change

REVISED

Mitigation strategies to cut global emissions of greenhouse gases

Mitigation refers to any method used to reduce or prevent the emission of greenhouse gases. Table 4.7 summarises the main strategies.

Table 4.7 Mitigation strategies to cut global emissions of GHGs

Strategy	Description
Energy efficiency and conservation	● New buildings regulations to improve heat insulation ● Financial incentives provided by governments, local authorities and energy companies to households for improvement to insulation
Fuel shifts and low-carbon energy sources	● Most governments are working towards an energy mix that reduces dependency on fossil fuels and increases the contribution from renewables
Carbon capture and storage	● Carbon capture and storage (CCS) refers to the technology that extracts CO_2 emitted by coal-burning power stations and transfers it to long-term underground storage ● There are three parts to the system: capturing, transporting and storing ● Cost and availability of suitable sites have been limiting factors
Forestry strategies	● Forests reduce CO_2 emissions to the atmosphere by storing carbon above and below ground; they also absorb CO_2 from the atmosphere ● Forest protection, reforestation and agroforestry are all important strategies
Geoengineering	**Geoengineering** projects include: ● increasing reflection of incoming solar radiation through reflecting plates in space or atmospheric aerosols ● removing CO_2 from the atmosphere through ocean fertilisation to stimulate the growth of phytoplankton which absorb CO_2 or by **enhanced weathering**

Geoengineering is the deliberate intervention in the Earth's climatic system with the aim of limiting the adverse effects of global warming.

Enhanced weathering is the artificial crushing of rocks to increase surface areas and absorption of atmospheric carbon dioxide by silicates to form carbonate minerals.

Revision activity

Choose one country example to describe and explain how government policy aims to conserve energy and improve efficiency.

Exam tip

Remember that there are costs and benefits to all mitigation strategies.

Now test yourself

TESTED

11 Why is it so important to protect trees?
12 How can ocean fertilisation reduce carbon dioxide in the atmosphere?

Answers on p. 221

Adaptation strategies to reduce the vulnerability of human populations at risk

Framework of adaptation

Adaptation anticipates the adverse effects of climate change and takes action to prevent or minimise damage. There are three main categories, summarised in Figure 4.7.

Retreat strategies	• Includes 'managed realignment' where vulnerable coastal communities are moved inland • Land use zoning prevents high-cost land uses such as housing and business from locating in vulnerable flooding area, e.g. on floodplains • The above strategies are adopted in countries with a high level of financial resources; in LIDCs, for poor communities in vulnerable locations, there may be few alternatives
Accommodation strategies	• Agriculture in particular will have to accommodate the impacts of climate change • Technological advancements are being made to help accommodate change, e.g. new crop strains, more efficient irrigation methods, replacement of cereal cultivation by livestock and tree farming, techniques to conserve soil moisture and drought-resistant crops • Adaptations to accommodate change within the water industry: reducing leakage, recycling water, increasing reservoir capacity and desalinisation of sea water • Climate change may lead to more hazardous and extreme weather events; accommodation includes education and increased preparation
Protection strategies	• Improved sea defences through hard engineering strategies such as sea walls and storm surge barriers • Soft engineering defences such as conservation of beaches, natural storage of water in river catchments, afforestation • Protection strategies will also be needed to address the threat to human health from climate change, e.g. protection from vector-borne diseases such as malaria

Figure 4.7 Framework of adaptation

Future adaptations to buildings, cities, transport and economies

- **Buildings:** adaptive strategies include insulation, increasing albedo with reflective roofing, night-time cooling systems, reducing high levels of glazing which trap heat in buildings, sunshades over windows.
- **Cities:** strategies include creating green infrastructure to reduce the urban heat island effect, greening cities to increase evapotranspiration, rooftop gardens to reduce albedo.

- **Transport:** there could be strategies to protect low-lying parts of rail networks from flooding, improve maintenance and protect road and rail networks from extremes of cold and heat. Severe weather may disrupt flights in the future; maintenance of runways, time separations of flights and new routes may all become future adaptations.
- **Economies:** how economies adapt to climate change in the future is difficult to predict and will vary across a range of countries at different stages of the development continuum. ACs will be in a better position to adapt to changes; LIDCs in the tropics and sub-tropics where temperatures may rise by 2°C or more will find it harder to adapt, particularly with a dependence on activities such as agriculture. Economic impacts through rising food prices, inflation, unemployment and declining exports may require adaptations to farming techniques, diet and sources of employment.

Case studies of two contrasting countries at different stages of economic development to illustrate mitigation and adaptation strategies

Revision activity

Make your own summary case study notes on the examples studied in class for this part of the course. Remember to focus on:

- current socio-economic and environmental impacts and the opportunities and threats they present
- technological, socio-economic and political challenges associated with effective mitigation and adaptation

Can an international response to climate change ever work?

Effective implementation depends on policies and cooperation at all scales

REVISED

Geopolitics associated with human responses to climate change

The role of the IPCC

IPCC reports are policy-relevant not policy-prescriptive; that is, they should inform policy makers. Five reports have been delivered; the latest in 2013 was used at the UN Climate Change Conference in Paris in 2015. Criticisms of the IPCC are summarised in Table 4.3 (see p. 132).

Exam tip

Climate change is a global problem affecting all countries in terms of cause and impact. However, in terms of responding, individual countries will always look after their own national interests.

International directives

Examples include the Kyoto Protocol (1997–2012), which was the first legally binding international agreement, driven by EU countries. A voluntary second commitment period from 2013 to 2020 has been supported by many ACs but has not been ratified by the largest carbon-emitting countries: the USA, China and India.

Carbon trading and carbon credits

A **carbon credit** is a certificate or permit that represents the right to emit 1 tonne of carbon dioxide; credits can be traded for money.

Carbon trading is a market-based system aimed at reducing GHG emissions. Richer countries can cut their emissions by paying for the development of a carbon-lowering scheme in a developing country. However, the effectiveness of these schemes has been criticised, with research indicating that emissions have increased.

A capping system has also been used to set an overall limit or 'cap' on the amount of emissions that are allowed from significant sources of carbon, e.g. power stations. Governments issue permits up to the agreed limit. If a company reduces its carbon emissions significantly, it can trade the excess permits on the carbon market for cash. If it cannot limit emissions it may have to buy extra permits.

The role of national and sub-national policy

Table 4.8 shows an example of a sub-national policy — that of California, USA. The state has become a world leader in implementing adaptation and mitigation policies to tackle climate change.

Table 4.8 California's climate change mitigation strategy

Fossil fuel-generated electricity	Retirement of power stations fuelled by fossil fuels in the long term. No new coal-fired plants will be built. Currently about 60% of electricity is produced in gas-fired power stations
Cap-and-trade	State scheme introduced in 2013. A market-based system to reduce carbon emissions. California's scheme is already the second largest in the world after the EU Emissions Trading System (EUETS). It covers 85% of carbon polluters in the state and despite strong economic growth in 2013–2015 is proving a success
Vehicle emissions	Stringent laws introduced to limit carbon emissions. The aim is to have 15% of cars powered by electricity, compressed natural gas and hydrogen fuel cells (as well as hybrids) by 2020. Currently nearly 3% of vehicles are powered by batteries
Renewables	California aims to generate 33% of its electricity from renewables by 2020 and 50% by 2050. Subsidies are available for renewables. 1940 MW of new solar energy was installed 2012–2015

Exam tip

You must make sure that you can provide details of a range of responses to climate change at a variety of **scales** — international, national and sub-national.

Now test yourself

TESTED

13 Why is climate change a transnational issue?
14 What are:
(a) the Kyoto Protocol
(b) carbon trading
(c) carbon credits

Answers on pp. 221–2

Summary

- Present-day climate change should be understood within the context of long-term climate change.
- The influence of natural forcing in the geological past should be understood.
- A new era, the Anthropocene, has linked climate change to human activity and, in particular, the process of industrialisation.
- There is a range of evidence to support climate change since the late nineteenth century.
- Anthropogenic greenhouse gas emissions have increased but the pattern varies spatially.
- There is considerable debate over climate change driven by a range of agendas.
- The impacts of climate change are global and dynamic.
- A range of mitigation strategies exist from global to local levels.
- Effective responses require cooperation at all scales and geopolitics is a key influence.

Exam practice

1 Explain how carbon capture and storage can cut greenhouse gas emissions. [4]

Table 4.9 Yearly number of extreme weather events in European Economic Area (EEA) member and collaborating countries in the period 1996–2010 (European Environment Agency, 2016)

	1996	1997	1998	1999	2000	2001	2002	2003	2004	2005	2006	2007	2008	2009	2010
Extreme weather events	19	29	31	45	64	40	60	45	38	90	46	63	29	43	59

2 Using the extreme weather event data in Table 4.9, calculate the interquartile range (IQR). You must show your working. [4]
3 Using evidence from Table 4.9, analyse the changes in extreme weather events. [6]
4 Examine the interdependence between governments and the role of international organisations in responding to climate change. [8]
5 'Adaptation strategies provide the best chance of reducing the risks from climate change.' To what extent do you agree with this statement? [12]

Answers and quick quiz 4A online

ONLINE

Option B Disease dilemmas

What are the global patterns of disease and can factors be identified that determine these?

Diseases can be classified and their patterns mapped. The spread of disease is complex and influenced by a number of factors

REVISED

How diseases can be classified

Diseases can be classified in a number of ways. Table 4.10 summarises the main ones.

Table 4.10 Disease classifications

Infectious	Spread by pathogens such as bacteria, viruses and parasites, e.g. malaria. Most can be transmitted from one person to another
Non-infectious	These are not communicable
Communicable	Infectious diseases which spread from one person to another but do not require quarantine
Non-communicable	These are not spread from person to person but have causes related to lifestyle, e.g. heart disease, some cancers; to nutritional deficiencies, e.g. rickets; or to genetic inheritance, e.g. heart disease
Contagious	A class of infectious disease easily spread by direct or indirect contact between people, e.g. typhoid, Ebola
Non-contagious	These cannot be spread by contact between people
Epidemic	A disease outbreak that attacks many people at the same time and spreads through a population in a restricted geographical area
Endemic	Exists permanently in a geographical area or population, e.g. sleeping sickness in rural areas of sub-Saharan Africa
Pandemic	An epidemic that spreads globally, e.g. Asian flu (1957–1958)
Zoonotic	Infectious diseases transmitted from animals to humans, e.g. rabies, plague

Typical mistake

Read questions carefully regarding the type of disease that you are being asked about, particularly if it is communicable or non-communicable. A mistake will significantly reduce your marks.

Exam tip

For this option, it is essential that you have a clear understanding of and accurate use of key terms.

Patterns of disease

Non-infectious disease is the main cause of death in developed areas but, increasingly, developing countries are affected. The pattern of other diseases may be determined by physical factors such as temperature, e.g. malaria.

Revision activity

Complete Table 4.11 or a diagram that summarises the global pattern of several key diseases, e.g. malaria, diabetes, cardiovascular disease. Include the name of the disease, a brief outline of its characteristics and an overview of its global distribution. An example has been done for you in Table 4.11.

Table 4.11 The global pattern of diseases

Malaria	An infectious, non-contagious tropical disease	Global distribution of malaria is concentrated in central and southern Africa; Latin America: north and western Brazil and Colombia, for example; and Southeast Asia: India, Pakistan and Indonesia, for example

Disease diffusion and spread

The spread of a disease is known as **disease diffusion**. There are several types, as shown in Figure 4.8.

- **Expansion diffusion:** the disease has a source and spreads outwards into new areas. The source area remains infected.
- **Relocation diffusion:** the disease leaves the area of origin and moves into new areas.
- **Contagious diffusion:** the spread is through direct contact with a carrier.
- **Hierarchical diffusion:** disease spreads through a structured sequence of locations, usually from large, well-connected centres to smaller, more isolated centres.

Disease diffusion is the process by which a disease spreads outwards beyond its geographical source.

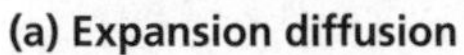

(a) Expansion diffusion

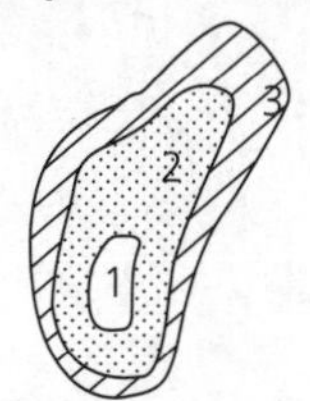

(b) Relocation diffusion

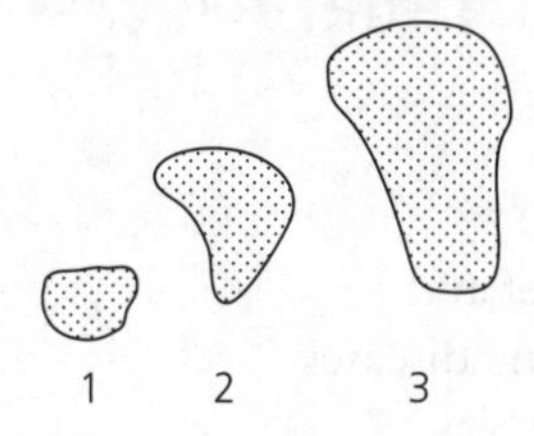

(c) Contagious diffusion

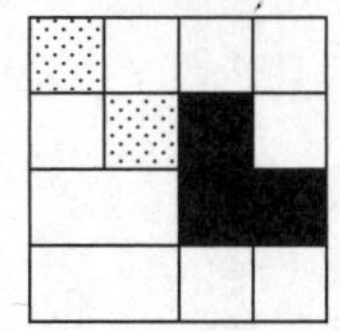

(d) Hierarchical diffusion

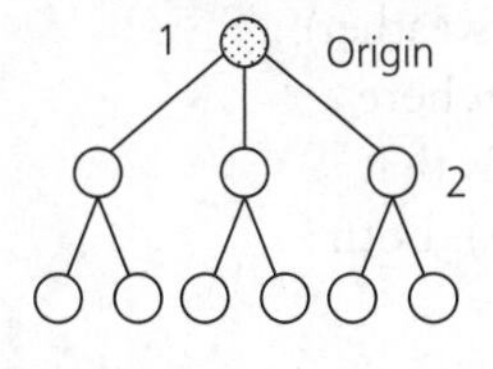

Figure 4.8 Types of diffusion

Hägerstrand's diffusion model

This model has been applied to the contagious diffusion of disease. The most important concepts of the model are:

- a neighbourhood effect: the probability of contact between those infected and those not infected and the influence of distance decay
- the number of people infected: represented by an S-shaped curve over time (Figure 4.9)
- the interruption of the progress of a disease by physical barriers

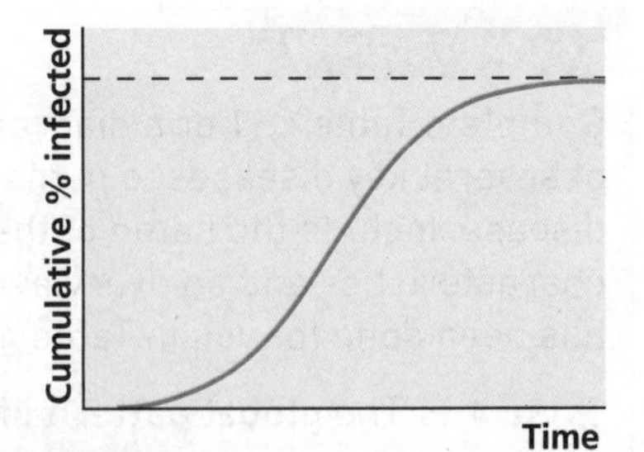

Figure 4.9 S-shaped logistic curve

Exam tip

A simple diagram can often replace lengthy description when referring to models and processes.

Phases of diffusion in Hägerstrand's model

- **Primary stage:** there is a strong contrast in disease incidence between the area of outbreak and remote areas.
- **Diffusion stage:** new centres of disease outbreak occur at distance from the source and this reduces the spatial contrasts of the primary stage.
- **Condensing stage:** the number of new cases is more equal in all locations, irrespective of distance from the source.
- **Saturation stage:** diffusion decelerates as the incidence of the disease reaches its peak.

Physical barriers: e.g. distance, mountains, sea, deserts, climatic change or general climatic conditions such as tropical or continental.

Socio-economic barriers: e.g. political border checks, imposition of curfews or quarantining. Most of these are put in place by international organisations and/or governments.

Now test yourself

TESTED

15 What is the difference between an infectious and a contagious disease?
16 Define hierarchical disease diffusion.
17 What are the key phases of Hägerstrand's diffusion model?

Answers on p. 222

There is a relationship between physical factors and the prevalence of disease which can change over time

REVISED

Global patterns of temperature, precipitation, relief and water sources and how these affect patterns of disease

Physical factors such as temperature, precipitation (i.e. climate), relief and water sources are important drivers of many diseases — vector-borne diseases such as malaria (absent from cool, dry climates) but also illnesses such as influenza, illness derived from vitamin D deficiency (related to low levels of sunlight) and respiratory illness (related to poor air quality and cold weather).

- **Climate (temperature and precipitation):** climate affects where disease vectors live. The most frequently referred-to example is that of mosquitoes and tsetse flies (which transmit sleeping sickness): both favour warm, humid, tropical conditions.

- **Relief:** altitude leads to falling temperatures and increasing rainfall. In parts of Africa, at higher altitudes it is too cold for mosquitoes. Flat floodplains can lead to diseases during periods of flood in low-income countries, e.g. hepatitis A and E and dysentery.
- **Water sources:** stagnant water affects the prevalence of water-borne diseases; parasitic worms in infected water spread diseases.

Exam tip

Pattern is an extremely important concept in geography. Practise skills in describing patterns and linking patterns.

Physical factors can influence vectors of disease

Examples of **vectors** include mosquitoes, tsetse flies, fleas, parasitic worms and some snails. Vector-borne diseases include dengue fever and African sleeping sickness. Malaria is the most widespread vector-borne disease, and is affected by the following physical factors:

- rainfall (puddles at the end of the rainy season)
- temperature (particularly active in places where average temperatures are between 18°C and 40°C)
- humidity (where average monthly relative humidity is over 60%)

A **vector** is a carrier of an infectious disease, e.g. mosquitoes carrying malaria.

How seasonal variations influence disease outbreaks

- Epidemics of influenza and respiratory illness peak in the winter months in the Northern Hemisphere.
- Vector-borne diseases transmitted by mosquitoes, flies and ticks peak during the rainy season in the tropics and sub-tropics.
- Fly populations are highest in South Asia in the pre-monsoon (March–April) and end of monsoon (September–October) seasons.
- The tsetse fly, which transmits sleeping sickness in west and central Africa, can live longer in the wet season.
- Freshwater snails, which transmit bilharzia to humans, follow a seasonal lifecycle pattern linked to precipitation and temperatures of 10–30°C.

Now test yourself TESTED

18 State **three** ways in which climate can influence patterns of disease.

19 How can relief effect patterns of disease?

20 Explain **one** seasonal influence on disease outbreaks.

Answers on p. 222

Climate change provides conditions for emerging infectious disease to spread to new places

Warmer and wetter conditions mean that vector-borne diseases such as malaria and dengue fever will spread in terms of their geographical area. Climate change is also responsible for the spread of Lyme disease and trypanosomiasis (sleeping sickness) as both diseases spread northwards. As temperatures rise, the World Health Organization (WHO) predicts that sleeping sickness will spread to southern Africa and affect 77 million people by 2090. A further consideration is that such diseases may disappear from some areas that become too hot, e.g. East Africa.

The conditions for zoonotic infectious diseases to establish and spread from animals to humans

Zoonotic diseases spread from animals to humans by viruses, bacteria, parasites and fungi. They include malaria, dengue fever and sleeping sickness. Dogs, bats, foxes and raccoons can all transmit rabies; poultry can transmit Asian flu. Probability of the spread of zoonotic diseases increases where:

- there is free movement of infected animals
- urbanisation creates habitats for animals, e.g. foxes, and brings them into closer contact with human populations
- there is not an effective vaccination programme for pets and livestock
- hygiene and sanitation are poor
- there is prolonged close contact between humans and animals, e.g. poultry farms

The WHO defines **zoonotic diseases** as diseases that are naturally transmitted between vertebrate animals and humans. A zoonotic agent may be bacteria, viruses or fungi.

Natural hazards can influence the outbreak and spread of disease

REVISED

Case study of **one** country which has experienced a natural hazard and the implications this has for a named disease

Revision activity

Choose an appropriate structure and make revision notes on a case study covered in class for this part of the course. Remember to focus on the following:

- geographical area covered by the hazard
- environmental factors affecting the spread of the disease
- human factors affecting the spread of the disease
- impacts of the disease on the resident population
- strategies used to minimise the impacts of the disease at national and international scales

Is there a link between disease and levels of economic development?

As countries develop economically the frequency of communicable diseases decreases, while the prevalence of non-communicable diseases rises

REVISED

Exam tip

Remember that developed countries can be subject to sudden 'shocks' as a result of the global spread of a communicable disease, but they have the resources in place to quickly educate people on how it can spread and to develop medication.

How rising standards of living influence a country's epidemiological transition

Economic and social development and the epidemiological transition

As countries develop, the health and wellbeing of the nation should also improve as more money is available to spend on agriculture, health services and basic infrastructure.

Economic developments

- Investment in agriculture to raise yields and farming efficiency in order to provide adequate good-quality food.
- Improved infrastructure so that food can be stored and distributed efficiently and basic services such as energy, sewage and clean water can reach the whole population.
- Investment in the health service.

Social developments

- Better education on sanitation, healthy diet and the spread of disease.
- Advances in medical care and availability of basic medicines and vaccinations.
- Better education and more opportunities to become fully trained health care professionals.
- Reduced infant mortality rates.

Epidemiological transition

A model suggests that over time as a country develops there will be a transition from infectious diseases to chronic and degenerative diseases as the main cause of death. The model, based on a line graph (Figure 4.10), was put forward by Abdel Omran in 1971. It can be seen as a sub-section of the Demographic Transition Model in the stages where medical advances impact birth and death rates.

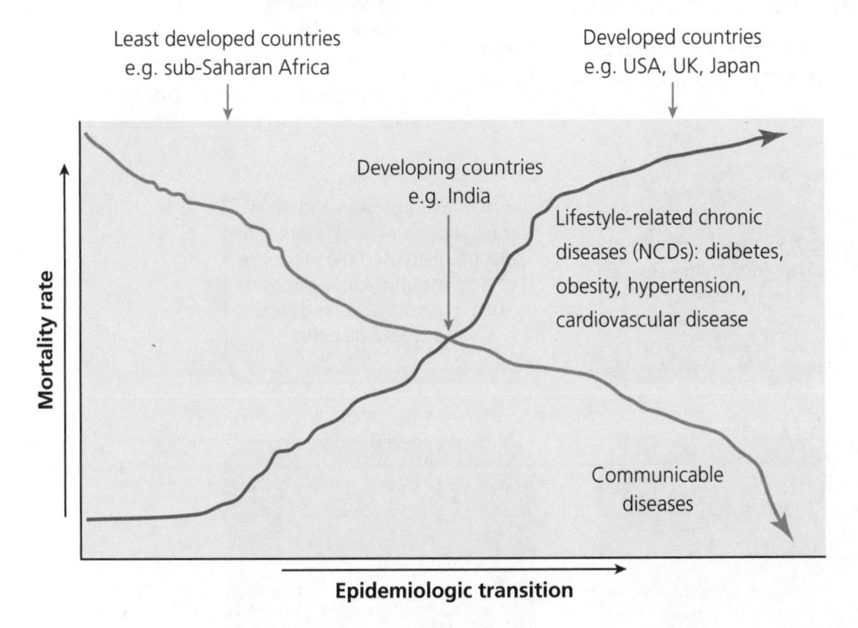

Figure 4.10 **Line graph outlining epidemiological transition**

The model is divided into three stages, with a fourth added in the 1980s (Table 4.12).

Table 4.12 The four stages of epidemiological transition

Phase	Life expectancy (years)	Change in socio-economic conditions	Causes of morbidity and mortality
Age of infection and famine	20–40	Poor sanitation and hygiene; unreliable food supply	Infections; nutritional deficiencies
Age of reducing pandemics	30–50	Improved sanitation; better diet	Reduced number of infections; increases in occurrence of strokes and heart disease
Age of degenerative and man-made diseases	50–60	Increased ageing; lifestyles associated with poor diet, less activity and addictions	High blood pressure, obesity, diabetes, smoking-related cancers, strokes, heart disease and pulmonary vascular disease
Age of delayed degenerative diseases	*c.* 70+	Reduced risk behaviours in the population; health promotion and new treatments	Heart disease, strokes and cancers are main causes of mortality but treatment extends life. Dementia and ageing diseases start to appear more

Because of variations in the pattern and pace of the transition, Omran identified three contexts to the model:

- **classical/western model:** e.g. western Europe where a slow decline in death rate is followed by lower fertility
- **accelerated model:** e.g. parts of Latin America where falls in mortality are much more rapid
- **contemporary/delayed model:** e.g. sub-Saharan Africa where decreases in mortality are not accompanied by decline in fertility

Exam tip

Remember that models are best used as a starting point to explain reality. They may provide a useful reference point but do not spend too much time on descriptive detail. A simple diagram may help, e.g. Figure 4.10.

The reasons why LIDCs have a higher prevalence of communicable diseases and ACs have a higher prevalence of non-communicable diseases

Figure 4.11 summarises some of the reasons why LIDCs have a higher prevalence of communicable diseases and ACs have a higher prevalence of non-communicable diseases.

Exam tip

The ability to evaluate models is also important.

Communicable diseases have largely been eliminated in ACs through diagnoses, treatment, high standards of living, clean water and good nutrition

Prolonged life in ACs means that degenerative illnesses have increased

In ACs, overnutrition and excess consumption of sugar, fats and salts has increased the incidence of non-communicable diseases such as cardiovascular disease and type-2 diabetes

Overnutrition is becoming an increasing problem in EDCs as affluence increases and dietary choices change. Other non-communicable diseases such a cancer are now increasing in EDCsand LIDCs

Geography is also a factor in the prevalence of communicable diseases in LIDCs. Tropical and sub-tropical locations give rise to diseases such as malaria, dengue fever, sleeping sickness, yellow fever and Ebola

Communicable disease accounts for the majority of deaths in the poorest countries because of poverty, lack of resources, inadequate nutrition, water pollution, lack of sanitation and poor hygiene

Figure 4.11 Reasons for the prevalence of communicable and non-communicable diseases

Now test yourself

TESTED

21 What is the epidemiological transition?
22 What are the **three** contexts to the epidemiological transition model?

Answers on p. 222

Case study of **one** country experiencing air pollution and the impact this has on incidences of cancers. Include the global and national solutions in dealing with this

Revision activity

Make your own revision notes in a suitable format for this section of the course.

How effectively are communicable and non-communicable diseases dealt with?

Communicable diseases have causes and impacts with mitigation and response strategies with a varying level of success

REVISED

Case study of **one** communicable disease at a country scale, either an LIDC or EDC

Revision activity

Make a revision summary of the case study covered in class for this part of the course. Remember to focus on the following:

- environmental and human causes of the disease
- prevalence, incidence and patterns of the disease
- socio-economic impacts of the disease
- direct and indirect strategies used by government and international agencies to mitigate against the disease and respond to outbreaks

Non-communicable diseases have causes and impacts with mitigation and response strategies with varying levels of success

REVISED

Case study of **one** non-communicable disease at a country scale, either an AC or EDC

Revision activity

Make a revision summary of the case study covered in class for this part of the course. Remember to focus on the following:

- social, economic and cultural causes of the disease
- prevalence, incidence and patterns of the disease
- socio-economic impacts of the disease
- direct and indirect strategies used by government and international agencies to mitigate against the disease

Typical mistake

Do not assume that it is only ACs that have effective health care and treatments. There are some very good examples in LIDCs and EDCs.

How far can diseases be predicted and mitigated against?

Increasing global mobility impacts the diffusion of disease and the ability to respond to it

REVISED

The role of international organisations

A range of international organisations have roles in the prediction and mitigation of disease. These include the WHO, UNICEF and NGOs such as Médecins Sans Frontières.

Increasing global mobility has the effect of leading to wider disease diffusion because of greater international flows of people, but improvements in transport and the effect of a 'shrinking world' can also enable international organisations to respond rapidly to disease outbreaks.

The World Health Organization (WHO)

The directing and coordinating body for global health, established in 1948, the WHO operates within the UN in Geneva, and has 194 member states. Wide-ranging activities include:

- data collection
- providing leadership in health matters
- technical support
- research and monitoring

The WHO publishes World Health Statistics annually, which gives a country-by-country insight into health risks, mortality rates, spread of disease and government spending. The WHO funds many research projects, often in conjunction with other organisations. The WHO takes a leading role in responding to major disease outbreaks such as the Zika virus and malaria. The WHO also delivers emergency aid, e.g. following the Nepal earthquake in 2015.

Exam tip

Make sure that you are able to compare and contrast the different roles and effectiveness of different international organisations, e.g. compare the WHO with UNICEF and an NGO such as the International Red Cross.

Example of a disease outbreak at a global scale

Revision activity

Based on an example studied in class, make revision notes on a global disease outbreak. Remember to focus on its rate of spread and pattern of outbreak distribution. An annotated map would be an effective way to summarise this.

Case study of the role that **one** NGO has played in dealing with a disease outbreak within **one** country at national and local level

Online (see p. 3) you will find the outline of a case study on the British Red Cross. This will show you how you could structure notes on an example for this part of the course.

Exam tip

Case studies are important to support and illustrate points made in extended responses. Learn specific facts to enhance your answers.

Mitigation attempts to combat global pandemics and overcome physical barriers

REVISED

Physical barriers which have positive and negative effects on mitigation

Barriers of disease diffusion can be classified as natural or physical barriers and human barriers (socio-cultural: when a person's beliefs or culture prohibit certain interactions; or political: borders).

The most important physical barrier is that of distance decay. The further a place is away from the source of a disease, the lower the incidence of disease. Other physical barriers include remoteness, e.g. remote regions such as rural peripheries, also mountainous regions and regions of extreme climate. Mountains and oceans also act as major natural barriers to the spread of disease. Physical barriers can also limit the spread of disease by vectors such as mosquitoes, e.g. deserts.

Physical barriers can have positive and negative effects on mitigation, as summarised in Figure 4.12.

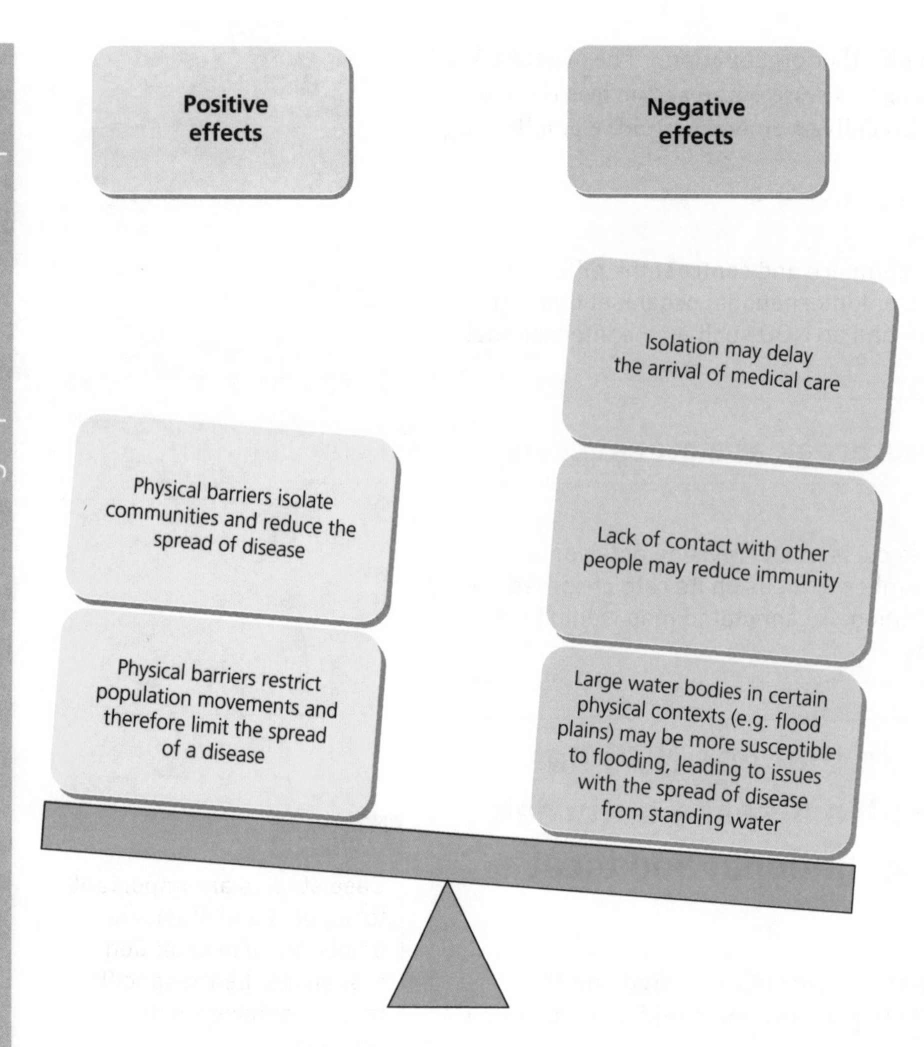

Figure 4.12 The positive and negative effects of physical barriers on disease mitigation

Now test yourself

TESTED

23 What is a disease barrier?
24 State **two** positive effects of physical barriers on the mitigation of diseases.

Answers on p. 222

Mitigation strategies used by government and international agencies to combat global pandemics

A pandemic is a worldwide **epidemic** caused by a virus that affects most or all age groups within months. In 2007, WHO in conjunction with FAO (the Food and Agriculture Organization) and the World Bank reviewed the status of H5N1 avian influenza in animals and assessed the related risks to human health following the emergence of a **pandemic** virus.

There were two aims: preventing the emergence and/or spread of a pandemic virus and preparing all countries to cope with a pandemic. The details of the strategic plan are shown in Table 4.13.

Epidemic: a disease outbreak that spreads quickly through the population.

Pandemic: an epidemic that spreads worldwide.

Table 4.13 Strategic action plan for H5N1 (avian flu) (adapted from WHO resource)

Strategic action	Aim
Reduce human exposure to H5N1 virus	Reduce opportunities for infection and reduce opportunities for pandemic virus to emerge
Strengthen the early warning system	Ensure that affected countries, WHO and the international community have all the data and specimens needed for an accurate risk assessment
Intensify rapid containment operation	Delay international spread of the virus
Build capacity to cope with pandemic	Ensure that all countries have formulated and tested pandemic response plans and that WHO is fully able to perform its leadership role
Coordinate global scientific research and development	Ensure that pandemic vaccines and antiviral drugs are rapidly and widely available shortly after the start of a pandemic and that scientific understanding of the virus evolves quickly

At the individual country (government) level, strategies may include providing guidance for members of the public and businesses on symptoms and on how to avoid catching and spreading the illness and what action to take on becoming unwell. Other strategies include: overseeing a coordinated response with health professionals, media, businesses, communities and voluntary organisations; and a prediction of possible socio-economic impacts.

Now test yourself

TESTED

25 What is a pandemic?

Answers on p. 222

Can diseases ever be fully eradicated?

Nature has provided medicines to treat disease for thousands of years

REVISED

Medicines from nature

Many modern medicines originate from wild plants and other natural sources such as microbes and animals (e.g. snake venom in drugs to treat heart conditions). This is not new — long before scientific medicine, plants were used for medicine, as reflected in their common names, e.g. woundwort and liver wort. Table 4.14 shows some medical drugs that are derived from natural compounds.

Table 4.14 Some medical drugs derived from natural compounds

Drug	Source	Growing conditions	Medical usage
Caffeine	Tea, coffee, cocoa and other plants	Tropical and sub-tropical conditions, temperatures 20–27°C and rainfall 1,000–2,000 mm/yr	Stimulant for central nervous system, heart, muscles. Migraine, epidural, anaesthesia etc.
Quinine	Dried bark of cinchonas (evergreen trees)	Average temperatures above 20°C, humid conditions, rainfall average 200 mm over at least 8 months	Malaria
Morphine	Dried latex seed pods of several species of opium poppy	Warm, humid conditions, temperatures 30–38°C	Pain reliever

Case study of one medicinal plant

Online (see p. 3) you will find a case study of the rosy periwinkle. This will show how you could summarise your notes for a case study on this part of the course.

Conservation issues relating to the international trade in medicinal plants

Medicinal plants are mainly derived from wild populations, and many are now at risk of extinction. Reliance on medicine from wild plants is particularly high in the developing world. Figure 4.13 summarises some of the main conservation issues arising from the international trade in medicinal plants.

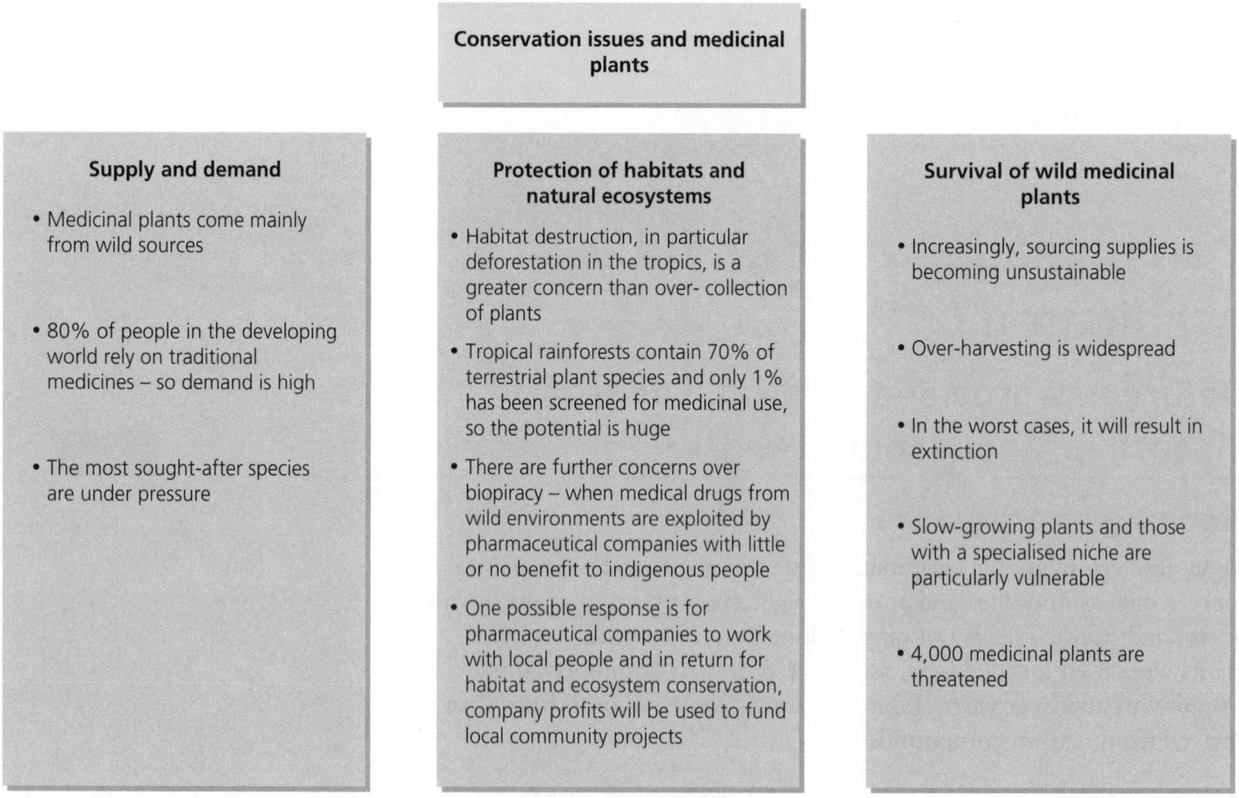

Figure 4.13 Conservation issues relating to international trade in medicinal plants

Top-down and bottom-up strategies that deal with disease risk and eradication

REVISED

Case study of the global impact of one pharmaceutical transnational

Revision activity

Make your own case study revision notes for the example studied in class for this part of the course. Remember to focus on the following:

- scientific breakthroughs made
- patents
- drug manufacturing
- global flows for distribution

Strategies for disease eradication at a range of scales

Global scale

Guinea worm (*Dracunculiasis*) is the first parasitic disease set for eradication. This is because diagnosis is easy (it relies on visual recognition of the emerging worm), the intermediate host is restricted to stagnant water bodies, interventions are simple and cost-effective, the disease has limited geographical distribution, and several countries in Africa, Asia and the Middle East have successfully eliminated the disease.

The process of eradication takes time: in 1981 WHO's decision-making body, the World Health Assembly (WHA), adopted a resolution supporting the opportunity to eliminate *Dracunculiasis*. In 2011, the WHA called on all member states to enforce nationwide surveillance to ensure eradication of *Dracunculiasis*.

The eradication strategy involves:

- mapping all endemic villages
- implementing effective case containment measures in all endemic villages
- implementing specific interventions such as ensuring access to safe water and health education
- reporting on a regular basis
- managing the certification process for global eradication country by country

Exam tip

Scale can be used to form an effective structure to extended responses.

National and local scale

Pakistan is one of only three countries in the world categorised as **endemic** by the Global Polio Eradication Initiative (GPEI), along with Afghanistan and Nigeria. Details of Pakistan's 2015 National Emergency Action Plan (NEAP) include the following:

- Emergency Operations Centres to be established at local level
- frontline health workers (mainly women) to be boosted by improved training and tools (including mobile phone apps), better communication systems and regular on-time payment

An **endemic** disease is one that exists permanently in a geographical area or a human group.

- vaccination to be made the norm for a community through the Emergency Operations Centres
- vaccine management including better refrigeration of vaccines
- improved health, sanitation and nutrition for the most vulnerable

The impact of grassroots strategies in educating communities and the role of women in combating disease risk

Grassroots strategies are usually small, community-based projects that focus on the needs of people and are often favoured by NGOs such as Oxfam. Education, assistance and engagement of local people are key features of such strategies.

The alternative to grassroots strategies is **top-down strategies** such as those led by international organisations and governments.

Women play a significant role in combating disease risk. Reasons include the following.

- They are often the primary carers for children and therefore are key to understanding and using vaccination and health awareness programmes.
- Increasingly, female frontline health workers have been seen to have a greater impact on education and vaccination programmes than male workers either because they can engage with and relate to women more effectively or because they understand a particular role undertaken by women that can spread disease, e.g. women sourcing clean drinking water and appreciating the need to filter drinking water to reduce the spread of Guinea worm.
- Women may have a specific role to play where a disease is passed on through pregnancy, e.g. Zika virus.
- In many societies women have the primary responsibility for maintaining hygiene in the home and for food preparation; poor practice in both allows disease to spread.

Exam tip

Make sure that you can relate the role of women in combating disease risk to a named example to illustrate the issue.

Now test yourself

TESTED

26 Why are certain medicinal plants at risk of extinction?
27 What does 'endemic' mean?
28 What is the difference between top-down and bottom-up strategies to deal with disease risk and eradication?
29 Why can female frontline health workers have a greater impact than men on eradication programmes?

Answers on p. 222

Summary

- Diseases can be classified into distinct types.
- Global patterns of disease vary and are affected by a range of factors.
- Physical factors impact on patterns and prevalence of disease and can change over time.
- Seasons, climate change and natural hazards are further influences on the outbreak and spread of disease.
- There is a relationship between disease and levels of economic development. The epidemiological transition model presents the theory of a decrease in the prevalence of communicable diseases and an increase in non-communicable diseases as a country develops.
- There are a range of mitigation and response strategies to deal with communicable and non-communicable diseases. These are implemented by national governments and international organisations.
- Global mobility impacts the diffusion of disease and responses to it at all scales from global to local.
- Physical barriers have both positive and negative effects on mitigation strategies and responses to global pandemics.
- Medicines from nature have treated diseases for thousands of years but increasingly they are threatened with extinction.
- Strategies for disease eradication include top-down and bottom-up approaches.
- Women have a key role in combating disease risk.

Exam practice

6 Explain how diseases can be classified as communicable and non-communicable. [4]

Table 4.15 Selected African countries, World Health Organization (2016)

Country	Estimated malaria deaths 2013 (thousands)
Nigeria	120
Democratic Republic of the Congo	50
Uganda	12
Mozambique	16
Burkina Faso	17
Ghana	15
Mali	20
Guinea	11
Niger	12
Malawi	7.8
Côte d'Ivoire	16
Cameroon	9.4
Kenya	9.9
United Republic of Tanzania	17

7 Using the data for estimated malaria deaths in Table 4.15, calculate the interquartile range (IQR). You must show your working. [4]

8 Using evidence from Table 4.15, analyse the contrasts in the number of malaria deaths. [6]

9 Examine how strategies used to mitigate non-communicable disease can impact social inequality. [8]

10 'Disease epidemic is inevitable following a natural hazard.' How far do you agree with this statement? [12]

Answers and quick quiz 4B online

ONLINE

Option C Exploring oceans

What are the main characteristics of oceans?

Oceans and seas make up about 71% of the Earth's surface, the deepest parts of the oceans make up about half of the Earth's surface and yet only about 1% of these regions has been fully explored.

The world's oceans are a distinctive feature of the Earth

REVISED

The global distribution of the world's oceans

Figure 4.14 shows the global distribution of the world's oceans.

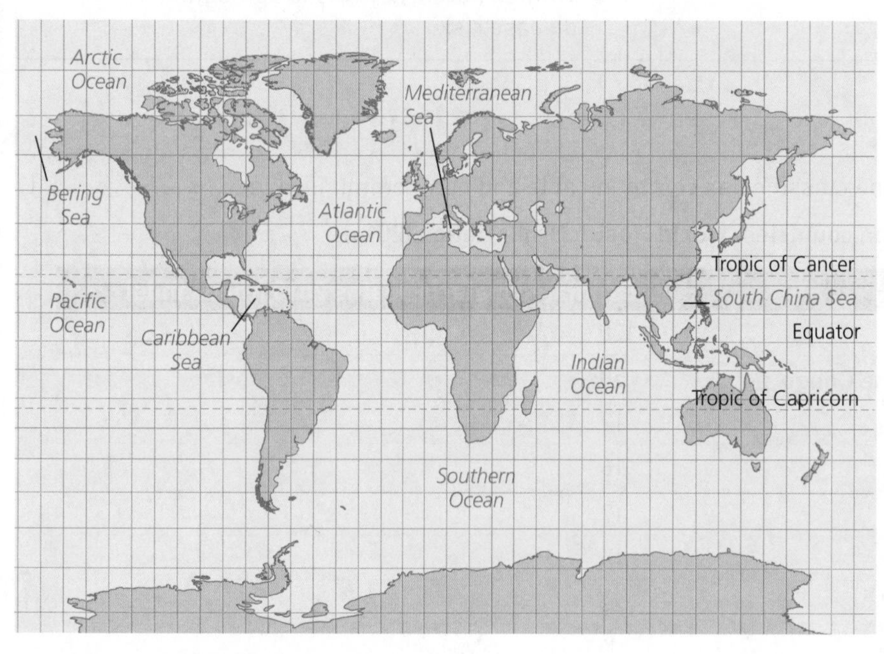

Figure 4.14 The world's oceans and seas

Exam tip

You should have a good working knowledge of the distribution of the world's major oceans and seas in addition to facts such as the size and average depth.

The relief of ocean basins

All oceans have a similar basic structure: edge of continent → water deepens across a **continental shelf** → at a **continental slope** the slope angle increases → the **continental rise** follows (a wide, gently sloping zone) → the deepest part of the ocean, the **abyssal plain**, is then reached → **guyots** are features of this plain (peaks that once rose above the ocean surface; erosion has flattened the top and reduced the height).

Some oceans, e.g. the Atlantic, have a chain of mountains marking mid-oceanic ridges and along the centre of the ridge is a **rift valley**. In other oceans, the margins are characterised by **subduction zones** where crustal plates are converging; a **trench** is a feature in these areas.

A **rift valley** is a valley formed by downfaulting between parallel faults.

A **subduction zone** is a zone where oceanic plate descends into the Earth's mantle and is destroyed.

An oceanic **trench** is a narrow, deep depression on the ocean floor adjacent to a subduction zone.

Revision activity

Annotate a copy of Figure 4.15, a simplified cross-section of an ocean basin, to explain its relief features. You can also include some named examples of features in a different coloured text.

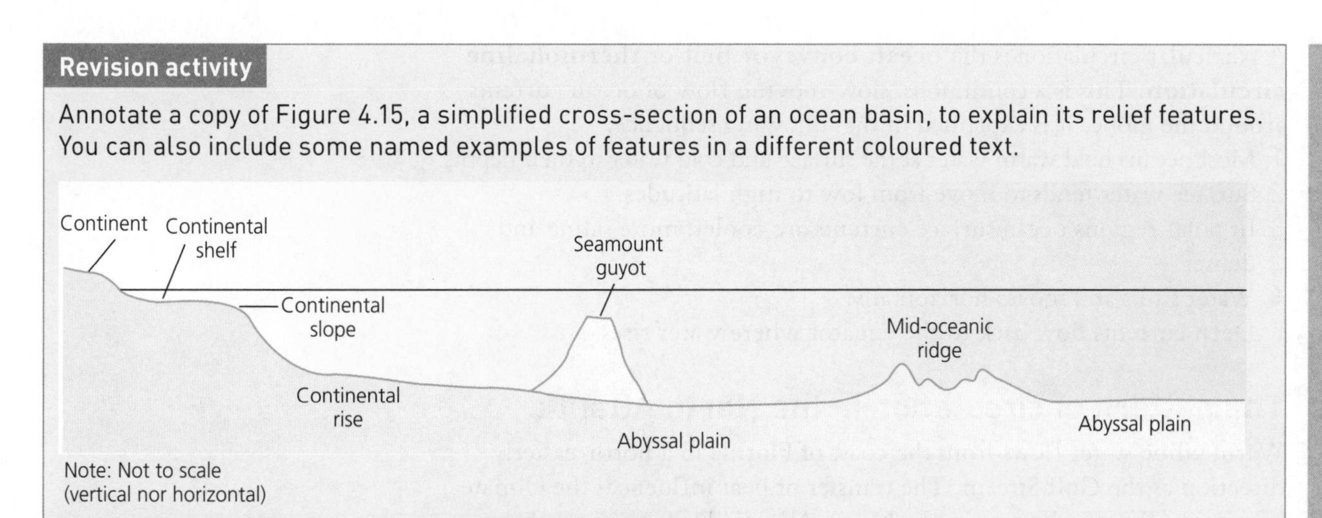

Figure 4.15 Simplified cross-section of an ocean basin

Exam tip

Practise simplified diagrams with clear annotation for this part of the course.

Water in the world's oceans varies

REVISED

Horizontal and vertical variations in salinity and temperature

Salinity

Salinity is a measure of the concentration of salt (sodium chloride). Fresh water is usually < 0.5 ppt (parts per thousand grammes of water); some oceans, e.g. the Pacific, can be 30–35 ppt. Salinity varies with depth and becomes lower at the surface (a gradient known as the halocline). Salinity has a wider significance as it influences water density; water density affects movements such as ocean currents which, in turn, affect global climate. Because salinity varies with depth, the different densities created affect the vertical movement of water in the oceans.

Temperature

High levels of energy are needed to heat water but, once heated, water retains heat very efficiently. The rise and fall of temperature on land surfaces is much more rapid and this contrast has an impact on weather and climate from global to local scales.

The global distribution of warm and cold surface currents

There are two types of ocean currents:

1. **Surface currents:** surface circulation — these waters make up about 10% of all the water in the ocean and form the upper 400 m of the ocean.
2. **Deep water currents:** thermohaline circulation — these waters make up the other 90% of the ocean and move around the ocean basins driven by forces of density (a function of salinity and temperature differences) and gravity. Ocean currents are influenced by several factors: solar heating, winds, gravity and the Coriolis force.

A particular circulation is the **ocean conveyor belt** or **thermohaline circulation**. This is a continuous, slow-moving flow of ocean currents around the globe. It is explained in the following sequence.

1. Most oceans hold warm water at the surface and cold water in their depths.
2. Surface water tends to move from low to high latitudes.
3. In polar regions ocean surface currents are cooler, more saline and denser.
4. Water sinks and moves horizontally.
5. Deep currents flow back to the equator where water rises.

The pattern of circulation in the North Atlantic

Warm saline water flows from the coast of Florida in a north-easterly direction as the Gulf Stream. The transfer of heat influences the climate of western Europe. Known as the North Atlantic Drift in its northern course, this warm current flows up to the Arctic. As the water cools, its density increases and it sinks. The water then travels back towards the equator where it will surface 150–250 years later.

In the northwest Atlantic, the cold Labrador Current flows south from the Arctic.

The North Atlantic also receives warm, salty water from the Mediterranean via the Straits of Gibraltar. Eventually its density matches that of the surrounding water and mixing occurs.

Now test yourself

TESTED

30 How does salinity affect ocean currents?
31 What are the **two** types of ocean currents?
32 Explain the sequence of thermohaline circulation.

Answers on p. 222

The biodiversity of oceans

REVISED

Factors affecting oceanic ecosystems

Light

Light is at its most intense level over the equator; with increasing distance from the equator, light is spread over a greater surface area because of lower angles. Light decreases with increasing depth of the oceans; therefore, photosynthesis is most active in the lighter layers, known as the **photic zone**.

Temperature

Sea surface temperatures are closely related to the amount of sunlight the sea receives. With increasing depth, temperature decreases.

Nutrient supply

Dissolved nutrients (e.g. nitrogen, iron and zinc) from rocks are carried to the oceans by rivers. Plankton hold these nutrients and they are passed through the food chain. Nutrients are returned to the water as waste from organisms or when dead organisms decompose. Nutrient levels are relatively low at the surface and may be quickly used up. Cold, dense, deep water tends to hold nutrients near the ocean floor. There are exceptions, e.g. the Southern Ocean and Antarctica, where a strong upward movement of deep water moves nutrients nearer the surface.

A comparison of inter-tidal and deep-water ecosystems

As a basic structure ocean ecosystems start with **producers** capable of trapping sunlight and converting it to chemical energy and organic matter through the process of photosynthesis. Phytoplankton are key ocean producers.

Biodiversity of the oceans varies with latitude and depth. A comparison can be made between the high **net primary productivity** of inter-tidal estuaries and shallow coastal waters where there are plenty of dissolved nutrients and the much less productive deep oceans such as the Antarctic.

A **producer** is an organism that captures energy from the sun during photosynthesis and stores this energy as organic matter.

Net primary productivity (NPP) is energy produced by plants, taking into account energy used for respiration.

An inter-tidal ecosystem: a salt marsh

A salt marsh is a vegetated mudflat or coastal wetland. They form in sheltered estuaries and places where shallow water covers a low coastal gradient (usually above 30° north and south of the equator). The ecosystem is affected by the fluctuations created by the incoming (flood) and outgoing (ebb) tides which cover and then expose the marsh.

Due to a combination of the tides, fluctuations in salinity and water-logged soils; salt marshes contain a variety of organisms. Green algae live on the mudflats and rooted plants such as glasswort grow on the marsh surface. Crustaceans and molluscs live on the mudflats and are fed on by wading birds such as curlew and oyster catchers. Goby fish inhabit the shallow creeks that cross the marsh surface. Phototrophic bacteria decompose plant matter which is then consumed by worms and molluscs.

Revision activity

Make a revision summary of the features of a **deep-water ecosystem** such as the Antarctic. Focus on points which make a comparison between this and an inter-tidal ecosystem, e.g. how physical conditions such as seasonal changes in sea ice greatly influence the ecosystem and how food chains of the Antarctic are relatively simple, and biodiversity is low.

Exam tip

Remember that comparison is not just two sets of descriptive facts. You need to be able to make clear, 'comparative' statements if required.

Revision activity

Draw a simple diagram to illustrate a food chain for (a) a named deep-water ecosystem and (b) a named inter-tidal ecosystem.

Now test yourself

TESTED

33 What is the photic zone?
34 How do nutrient levels vary within oceans?

Answers on p. 222

What are the opportunities and threats arising from the use of ocean resources?

Biological resources within oceans can be used in sustainable or unsustainable ways

REVISED

With advancing technology, humans can make more intensive use of biological resources within oceans. Fish stocks provide a **natural capital**, as they are not manufactured by humans and when they are caught and sold the yield is termed a **natural income**. An **ecosystem service** refers to the benefits people obtain from ecosystems.

Ecosystem services: the benefit people obtain from ecosystems also known as ecosystem goods and services.

Fish would be part of a **provisioning service**, e.g. food and water. There are also **regulatory services**, e.g. climate, **supporting services**, e.g. nutrient cycles, and **collateral services**, e.g. recreational benefits of ecosystems.

Case study of the management of one renewable biological resource within oceans

Revision activity

Make your own revision summary from a case study covered in class for this part of the course. Remember to focus on:

- the use and management of this resource
- the influence of stakeholders in its use and management
- the resilience of the resource

The use of ocean energy and mineral resources is a contested issue

REVISED

The use and management of ocean energy resources and sea-bed minerals

Figure 4.16 outlines the key features of the use and management of energy (non-renewable and renewable) and mineral resources in the oceans.

Typical mistake

Make sure your examples fit the category for which they are required, e.g. biological resource: fish; renewable ocean resource: waves.

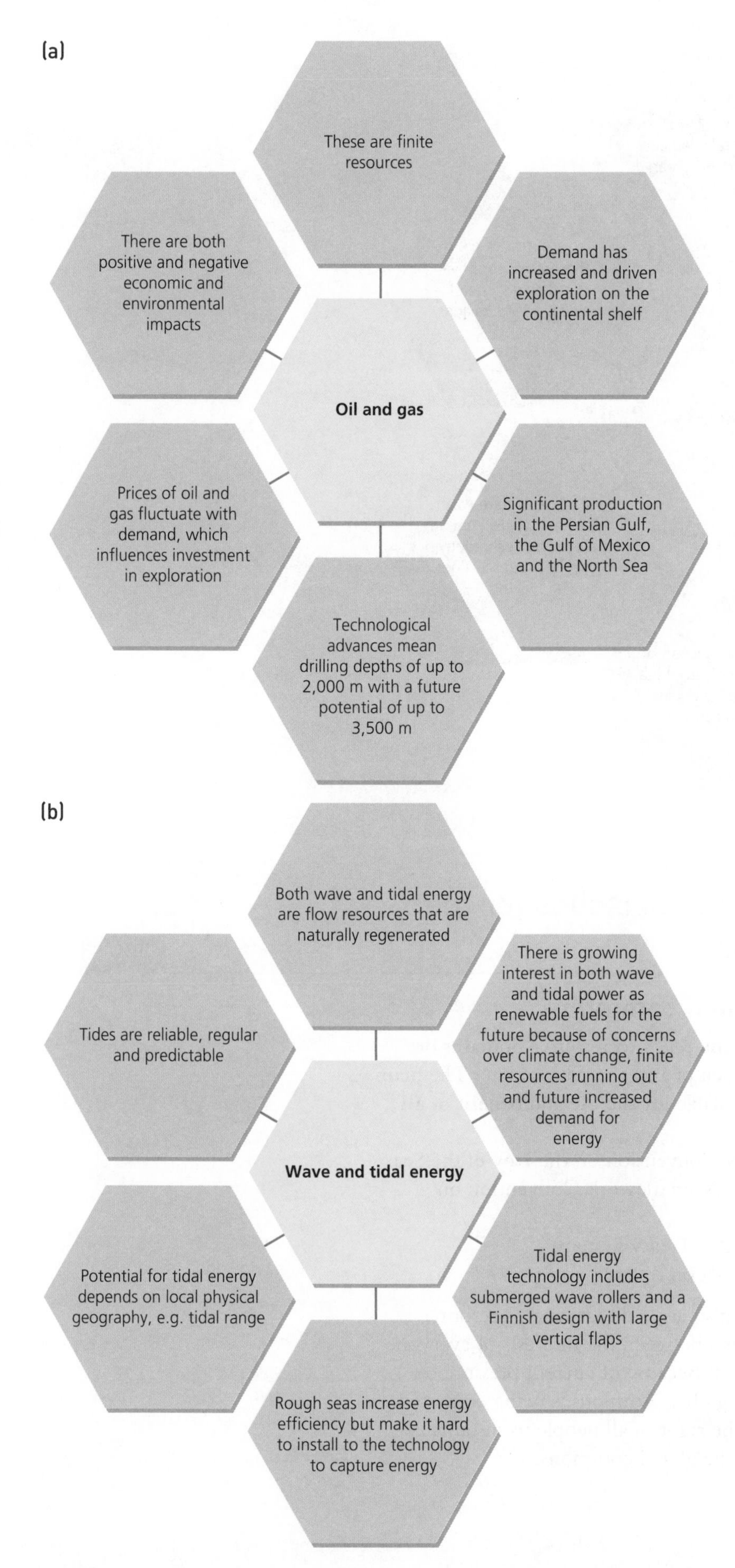

Figure 4.16 The use and management of oceanic energy and mineral resources. (a) Non-renewable energy resources; (b) renewable energy resources; (c) mineral resources

(c)

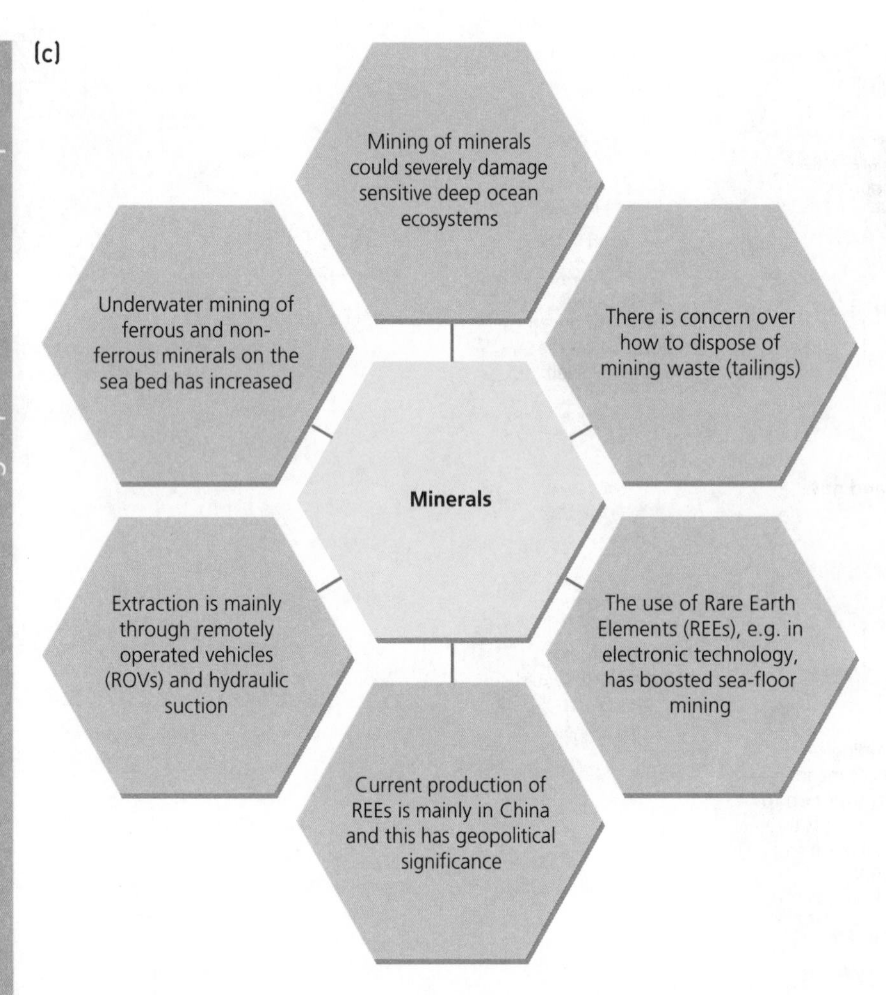

Figure 4.16 (continued)

Governing the oceans poses issues for the management of resources

REVISED

Oceans as part of the 'global commons'

The concept of the **global commons** refers to resource domains or areas that lie outside the political reach of any one nation state. The main concept is that the following are available for the use and benefit of **all** people:

- the high seas (covered by the UN Convention on the Law of the Sea)
- the atmosphere (covered by the UN Framework Convention on Climate Change)
- Antarctica (covered by the Antarctic Treaty System)
- outer space (covered by the 1979 Moon treaty)

Garrett Hardin (an American ecologist) argued that when a resource is seen as belonging to all, tension exists between the interests of everyone (the common good) and self-interests. Because of current pressure on scarce resources, the concept of the global commons is being contested. There is also pressure to maintain the right of all people to sustainable development and a need to protect the global commons.

The exploitation of ocean resources is a good example of the '**tragedy of the commons**', where people exploit a resource and take all they can, believing that the cost will be shared among a much larger number of people.

Oceanic management zones

The United Nations Convention on the Law of the Sea (UNCLOS) is an international agreement that defines the rights and responsibilities of countries with regard to the coastal zone and beyond. It is outlined in Figure 4.17.

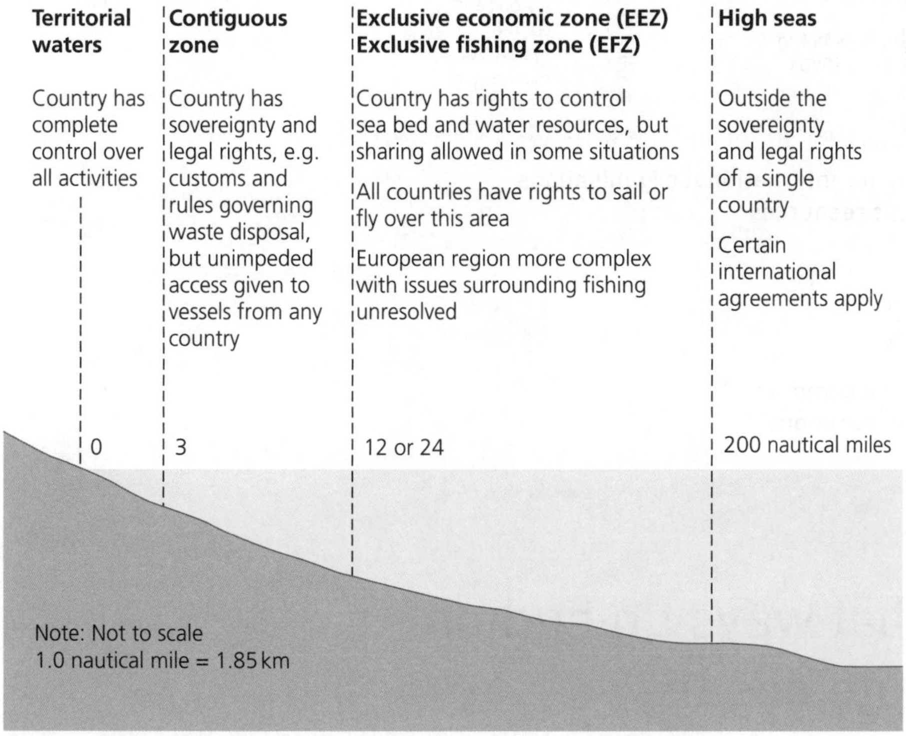

Territorial waters
Country has complete control over all activities

Contiguous zone
Country has sovereignty and legal rights, e.g. customs and rules governing waste disposal, but unimpeded access given to vessels from any country

Exclusive economic zone (EEZ)
Exclusive fishing zone (EFZ)
Country has rights to control sea bed and water resources, but sharing allowed in some situations
All countries have rights to sail or fly over this area
European region more complex with issues surrounding fishing unresolved

High seas
Outside the sovereignty and legal rights of a single country
Certain international agreements apply

0
3
12 or 24
200 nautical miles

Note: Not to scale
1.0 nautical mile = 1.85 km

Figure 4.17 Ocean management zones according to UNCLOS

Concerns relating to oceanic management zones include the following:

- The zones can be disputed and they are difficult to implement.
- If there is the prospect of access to valuable resources, disputes occur.
- There are other treaties governing waste and fishing, for example, and this creates confusion.
- Management issues that have arisen since UNCLOS include ocean acidification, fishing in deep oceans and **bio-prospecting**.

Bio-prospecting is the discovery and commercial exploitation of biological resources.

Resource management through international organisations

The **International Seabed Authority**, **Marine Reserves Coalition** and the **International Whaling Commission** are examples of other international attempts to oversee and manage the exploitation of oceanic resources.

Revision activity

Produce a diagram to summarise the features of the international organisations listed in bold in this section. Figure 4.18 shows an example of layout.

International Seabed Authority	International Whaling Commission	Marine Reserves Coalition
• Created by the UN • Oversees the exploitation of sea-bed resources	• Founded in 1946. • 88 member countries • Banned commercial whaling in 1986a	• Protects marine habitats • Recognises the biological, geological, historic and cultural features of marine locations

Figure 4.18 **Revision diagram for international organisations attempting to manage oceanic resources**

Now test yourself

TESTED

35 What is meant by 'the global commons'?
36 What is the 'tragedy of the commons'?

Answers on p. 222

How and in what ways do human activities pollute oceans?

There are a variety of pollutants that affect the ocean system

REVISED

Pollution is the damage caused when human activity adds harmful substances to the environment. The pollution may be **point sourced** from one identified location or **non-point sourced** from widespread origins. It was thought that vast oceans could dilute pollutants but now the scale and highly toxic nature of many pollutants are having a concerning adverse impact on marine ecosystems. Pollutants can travel long distances from their source via rivers or in the atmosphere. Pollutants of current concern include:

- plastics
- fossil fuel burning
- fertilisers
- pesticides
- industrial waste
- radioactive waste

The major sources of ocean pollutants

Revision activity

Based on specific examples studied in class, complete Table 4.16 to summarise the pollution concerns from combustion of fossil fuels and domestic and industrial pollutants (plastics, heavy metals, nuclear waste).

Table 4.16 Sources of ocean pollution

Combustion of fossil fuels at sea	Domestic and industrial pollutants, e.g. plastics, heavy metals or nuclear waste

Exam tip

Ocean pollution is a major environmental concern that is frequently reported in the news with new examples and updated statistics. Make sure that the information you use is current.

Off-shore oil production and transport poses threats for people and the environment

REVISED

Case study of one oil spill

Revision activity

Make your own case study summary for this part of the course from an example covered in class. Remember to focus on the following:

- impacts on the physical environment and marine ecosystems
- impact on human activities
- management of the spill and its impacts

The pattern of ocean currents can disperse and concentrate pollution

REVISED

How pollution can spread around the globe via oceanic circulation and its impact on marine ecosystems

Marine debris (plastics, metals, rubber etc.) enters oceans via rivers, from beaches and from dumping by ships at sea. It is a long-term, persistent problem, posing threats to ships and wildlife. One of the most common problems is plastics, which do not biodegrade. **Ocean gyres** are systems of oceanic circulation which transport and distribute pollution over vast distances and can lead to areas of accumulation.

Case study of the accumulation of plastic in one ocean gyre

Online (see p. 3) you will find a diagram showing how a case study of the Pacific could be summarised. You will need to add more detail.

How is climate change impacting the ocean system?

Climate change is altering the nature of the ocean's water

REVISED

The Earth's climate shows a pattern of variability. However, there is overwhelming evidence of current meaningful change that is directly affecting oceans. The scientific consensus is that increases in GHG emissions over the past 200 years are responsible for current global warming.

Two of the main impacts on oceans — **acidification** and **warming** — are summarised in revision diagrams in Figures 4.19 and 4.21. Figure 4.20 explains the process of acidification.

Ocean acidification

Oceans are a sink for carbon. As surface temperature of the oceans rises, their ability to absorb CO_2 reduces. Average global surface ocean pH has fallen from 8.2 to 8.1, still alkaline, but the process is known as acidification as this represents a 30% increase in acidity

Ocean acidification has the following impacts on marine life:

- Molluscs and crustacea at the bottom of the food chain are unable to build strong shells, so they are more susceptible to predators, numbers are reduced, and wildlife higher up the food chain suffer from a reduced diet
- Declining number of organisms with shells affects the carbon cycle, as when they die their shells form a store of carbon in rocks for millions of years
- Adaptations to more acidic oceans exist around underwater volcanoes
- Some of the most productive fishing areas exist where water upwells from the depths bringing CO_2 and nutrients to the surface

Ocean acidification has the following impacts on fish stocks:

- Early life stages of fish are vulnerable to acidic sea water
- Some predatory fish switch prey and as a result show few negative effects of acidification
- Other predatory species are in decline

Figure 4.19 The impacts of ocean acidification

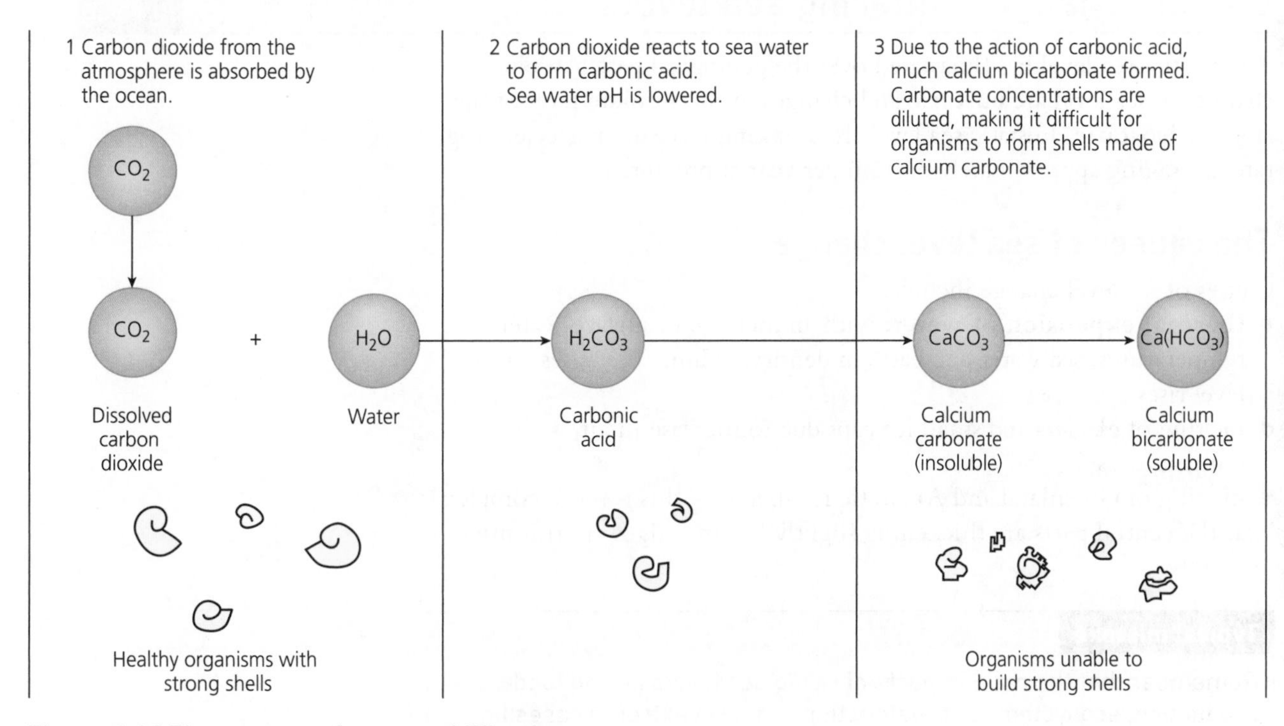

Figure 4.20 The process of ocean acidification

Rising temperatures

Warming oceans and coral reefs

- Corals are marine polyps which live in symbiosis with algae. The algae release nutrients by photosynthesis which the polyps feed on. The algae contain pigments which give coral its colour
- Coral require particular living conditions: ideally 26°C, salinity of 30–32,000 ppm, water depth of 25 m or less, light for the algae to photosynthesise, clear water and some wave action to oxygenate the water
- Warming oceans push corals to their thermal limits
- In extreme temperatures the relationship with the algae is lost and the coral lose their colour — coral bleaching occurs

Coral bleaching

- Coral bleaching is evident all around the tropics
- It is a complex process also associated with salinity
- Observations of entire coral communities bleaching have been increasing over the past 30 years
- Coral mortality has reached up to 80% in the Maldives and Seychelles
- Satellite measurements of increases in sea surface temperature can be used to predict bleaching events

Figure 4.21 The rising temperatures of oceans and threats to coral ecosystems

Exam tip

Make sure that you can refer to a named example of coral bleaching to illustrate the full impacts of this process.

Climate change is altering sea levels

REVISED

The Earth's sea level has fluctuated over the geological past. Absolute changes in sea level are **eustatic** and changes in the absolute level of the land are **isostatic**. Eustatic sea level rise is taking place at an accelerating rate, averaging approximately 3.0 mm per year at present.

The causes of sea level change

Causes of sea level change include:

- **thermal expansion of water:** with an increase in surface water temperatures, sea water decreases in density, volume increases and sea level rises
- melting of glaciers and small ice caps due to increase in air temperatures
- melting of Greenland and Antarctic ice sheets — this is more complex as the central parts are thickening slightly but the edges are thinning rapidly

Typical mistake

Remember that the main impacts of rising sea levels are on food production, ecosystems and water supply; only in extreme cases (in the short term) will rising sea levels result in 'washover'.

Case study of one island community

Revision activity

Make your own revision notes for a case study covered in class. Remember to focus on:

- threats to island communities
- the impact on communities
- adaptations by governments and communities in the short and long term

Climate change is altering high latitude oceans

REVISED

High latitude oceans are the Arctic and the Southern Ocean around Antarctica. Sea water freezes in these oceans at −2°C because of the salt content. There is very little heat input into the heat budget system as the sun's angle is low and heat energy is spread over a large area, and also large amounts of incoming solar radiation are reflected.

The impact of global warming on the extent of sea ice

Sea ice area

Between 1978 and 1996 the area of Arctic sea ice decreased by 2.9%; in 2015 ice covered 4.0 million km^2 compared with a long-term average of 15.0 million km^2. Figure 4.22 shows the interaction between the area of sea ice and warming of the Arctic.

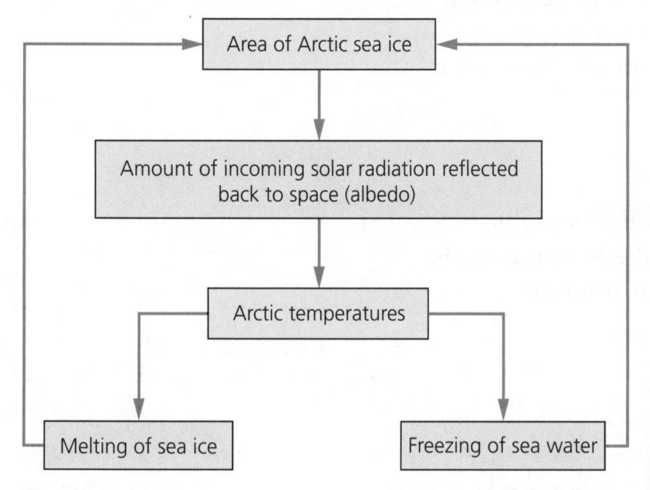

Figure 4.22 The interaction between area of sea ice and warming of the Arctic

Thickness of ice

In the 1960s Arctic sea ice averaged 4 m thick at the centre, whereas it is now about 1.25 m thick.

The main concern is that as sea ice reduces in extent, more solar radiation will be absorbed by the sea and not reflected, leading to further sea warming and more melting of sea ice.

Now test yourself

TESTED

37 What are ocean gyres?
38 State the main impacts of climate change on oceans.
39 What are eustatic and isostatic changes in sea level?
40 Name **three** causes of sea level change.

Answers on pp. 222–3

Case study of the Arctic region

Online (see p. 3) you will find a case study on the Arctic, illustrating the implications of change in sea ice. This is an example of how you could structure a summary of a case study for this part of the course. Place-specific facts and detail need to be added.

How have socio-economic and political factors influenced the use of oceans?

Oceans are vital elements of the process of globalisation

REVISED

Globalisation is the growing integration and interdependence of people's lives. It has economic, socio-cultural, political and environmental components. In relation to the oceans this means: more trading of goods across the oceans; more movement of people and resources (e.g. energy and minerals); more transport of people through tourism; and increasing interconnections, e.g. fibre-optic cable laying.

The pattern of principal shipping routes

The principal shipping routes follow an east–west corridor between North America and Europe and Pacific Asia through the Suez Canal, the Strait of Malacca and the Panama Canal, and a corridor from Europe to South America (Figure 4.23).

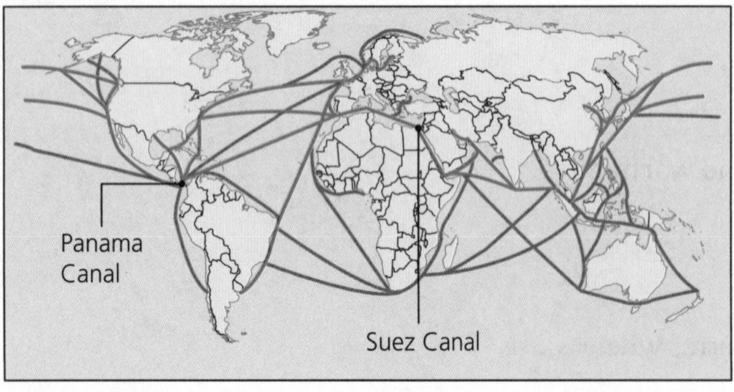

Key
— Core route — Secondary route

Figure 4.23 Early twenty-first century shipping routes

The direction and type of trade across the oceans

Figure 3.9, p. 93, shows the global pattern of inter-regional trade. Ocean trade connects the producers and consumers of raw materials and manufactured goods. It is a pattern reflecting the size and affluence of markets and is dominated by the east–west flow of goods across the Northern Hemisphere between the advanced economies of the EU, the USA, China, Japan and the Russian Federation.

It is important to consider the type of trade alongside its direction. Categories for the types of trade are agricultural, manufactured goods, energy and minerals (raw materials).

Revision activity

Annotate a world map to show the main trade flows in terms of the type of trade.

Oceans are important spaces where countries challenge each other

REVISED

Many marine boundaries continue to be disputed, particularly where there are known to be resources of energy, minerals or fish stocks, for example. Tensions are likely to increase as populations continue to grow and become more affluent, so increasing the need for resources as reserves that are currently exploited run out and as the Arctic Ocean becomes more accessible. Current geopolitics means that superpowers such as the USA and China seek to maintain and exert power via the oceans.

The use of oceans by countries to exert their influence

The South China Sea is a particularly good illustration of how countries seek to exert influence by use of the ocean. Figure 4.24 summarises this example (to be read from bottom to top).

- The region is also important to the USA strategically as a trading location worth US $1.2 trillion
- The USA is an ally of Taiwan and South Korea and will protect their security
- There is a fear that tension between the USA and China may grow over disputes in the South China Sea

Recent incidents include:

- 1974 and 1988 armed clashes between China and Vietnam
- 2012 China and the Philippines clash over the Scarborough Shoal
- 2012 China formally creates Sansha City in the Paracels
- 2014 China sets up a drilling rig near to the Paracels; collisions occur between China and Vietnam
- 2015 US satellite shows China building infrastructure on some of the Spratly Islandst

- Two island chains, the Paracels and the Spratlys, are claimed in whole or part by countries including China, the Philippines and Vietnam
- The attractions include reserves of oil and gas and also trade routes to China

Figure 4.24 Marine conflict in the South China Sea

Exam tip

When referring to case studies, factual detail is important. An annotated map is a good method of making revision notes and would suit the issue of marine conflict.

Oceans present hazardous obstacles to human activities

REVISED

Large numbers of refugees use oceans to escape conflict and find security. The Mediterranean is a particular 'hot spot' as migrants flee conflict in Syria, Yemen and parts of Africa. A further issue is piracy, which presents a modern-day threat to cargo ships and tourists.

Twenty-first century piracy

Piracy takes place out at sea and close to the shore, and in harbours and ports. There is a close geographical link to trading routes but it also occurs in specific areas such as the western Indian Ocean and Southeast Asia. Trade through the Suez Canal and the Malacca and Singapore Straits is also vulnerable.

Piracy is the act of boarding a vessel with the intent to commit theft or any other crime.

Modern-day piracy often involves the crew or tourists being held for ransom. There is also a clear seasonal pattern as a result of weather conditions, avoiding the strong winds of the summer monsoon season. As a response the EU, NATO and the USA, together with countries such as Russia, India and China, have pooled resources to effectively reduce the number of incidents. The threat remains because the causes of piracy — poverty, organised crime, dysfunctional governments and loss of traditional fishing areas — are complex and spread across several countries.

Now test yourself

TESTED

41 State **two** impacts of globalisation on oceans.
42 Give **two** reasons why some marine boundaries are disputed.
43 Give **two** causes of piracy in the twenty-first century.

Answers on p. 223

The use of oceans as escape routes for migrants

There are several types of migrant (Figure 4.25). People migrate because of a range of **push** (reasons for leaving) and **pull** (attractions of the destination) factors (Figure 4.26). Migration may also be **forced** when there is no other option.

Economic reasons are at the forefront of many moves. In areas of conflict such as the Middle East and parts of Africa, political and religious persecution leads desperate individuals to flee, risking their lives in difficult ocean crossings.

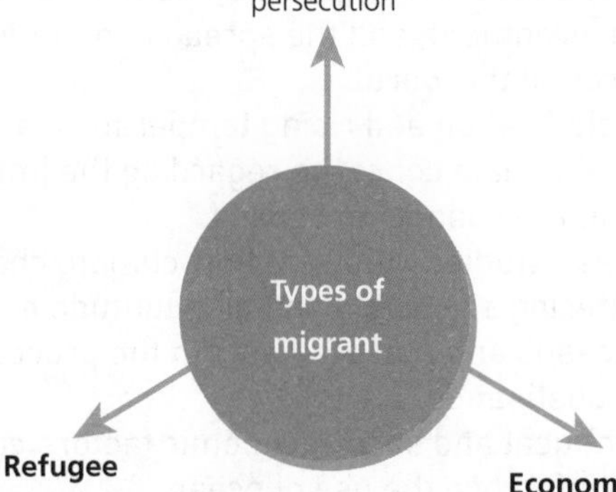

Figure 4.25 Types of migrant

Typical mistake

Migration does not include short-term movements such as tourism.

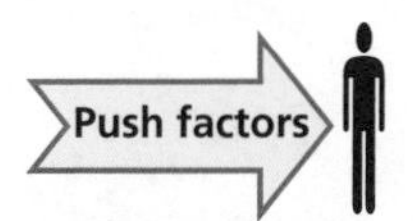

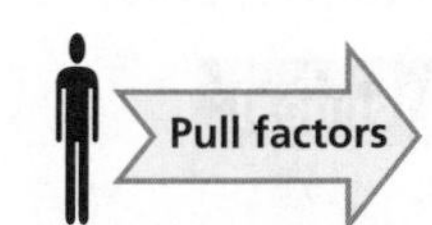

ORIGIN

'Forcing' factors

- War, conflict, political instability
- Ethnic and religious persecution
- Natural and man-made disasters such as earthquakes, tsunamis, drought, famines

Socio-economic conditions

- Unemployment, low wages or poor working conditions
- Shortage of food

DESTINATION

Associated with voluntary migration

- Better quality of life, standard of living
- Varied employment opportunities, higher wages
- Better healthcare and access to education services
- Political stability, more freedom
- Better life prospects

For retirees:

- Specific type of environment with a range of services to cater for their needs

Figure 4.26 Push and pull factors

Revision activity

Based on an example studied in class, make revision notes on one located example of migrants escaping via ocean routes. Include the background context (the country/countries of origin and the target host country/countries, some specific facts about numbers involved, the advantages and disadvantages of the move).

Summary

- Oceans have a range of distinctive features: underwater physical geography, salinity, temperature, nutrient supply, depth, biodiversity and ecosystems.
- There is a global system of warm and cold currents both at the surface and deep within oceans.
- Oceans provide a range of biological resources which are used in sustainable and unsustainable ways.
- The use of oceans' energy and mineral resources is a contested issue.
- The effective governance of oceans is a complex issue.
- Pollution of oceans is a major environmental concern. Issues include the variety of pollutants, the impacts on humans and the environment, and the spread of pollution around the world.
- Acidification and rising temperature are two of the main concerns regarding the impact of climate change on oceans.
- Case studies illustrate how climate change is altering sea levels and high latitude oceans.
- Oceans are vital elements in the process of globalisation.
- Political and socio-economic factors continue to influence the use of oceans.

Exam practice

11 Explain how nutrient supply varies vertically in the water column. [4]

Table 4.17 Fish landings in thousands of tonnes, 2014 (OECD, 2016)

Country	Fish landings (thousands of tonnes)
Greece	51
Estonia	52
Germany	57
Sweden	68
Portugal	130
Australia	152
Italy	177
New Zealand	289
Denmark	598
Argentina	786
Spain	1,211
Mexico	1,404
Norway	2,028
Japan	3,769

12 Identify **three** limitations of the data evidence in Table 4.17. [3]
13 Explain the impacts of an oil spill on the marine ecosystem. [6]
14 Examine how flows and connections across oceans are changing place profiles. [8]
15 'Changes in the ocean resulting from climate change will have the greatest impact on indigenous communities.' How far do you agree with this statement? [12]

Answers and quick quiz 4C online

ONLINE

What is food security and why is it of global significance?

There is a consensus that we can produce enough food. However, there is a global mismatch between demand and supply. The future of food will be determined by the efficient functioning of the global food system and the physical, economic, social and political factors that affect it.

The concept of food security is complex and patterns of food security vary spatially

REVISED

Defining what it means to be food secure

A commonly used definition of food security comes from the United Nation's Food and Agriculture Organization (FAO):

> Food security exists when all people, at all times, have physical and economic access to sufficient, safe and nutritious food that meets their dietary needs and food preferences for an active and healthy life.

The World Food Programme (WFP) identifies 'three pillars' of:

- **availability:** this addresses food supply
- **access:** this relates to household-level access to sufficient food
- **utilisation:** the intake of food must result in the body gaining sufficient nutrients and energy

The FAO also recognises '**stability**' as a fourth pillar to food security. This is important because food security is dynamic and can also be either long term or short term in nature.

Food security is a complex and contested term; discussion points include food distribution, future demand given population growth, and supply in light of climate change issues.

Now test yourself

TESTED

44 What is food security?
45 What are the **four** dimensions of food security?

Answers on p. 223

Current trends in global food security

A wide variety of data exists to measure food security. Examples of data sources include statistics on global hunger, undernourishment, daily calorie intake and per capita food consumption. Data from the World Health Organization on global and regional per capita food consumption show:

- there has been an increase in food consumption (kcal per capita per day) globally and in all categories apart from transition countries
- industrialised countries have the highest levels of consumption and sub-Saharan Africa the lowest (2015)

- developing countries, sub-Saharan Africa and South Asia fall below the global figure for 2015

The **Global Food Security Index** is a composite index providing a worldwide overview of countries most and least vulnerable to food insecurity. It combines 28 indicators which cover three key areas: food affordability, availability, quality and safety. The Global Food Security Index for 2015 is shown in Figure 4.27.

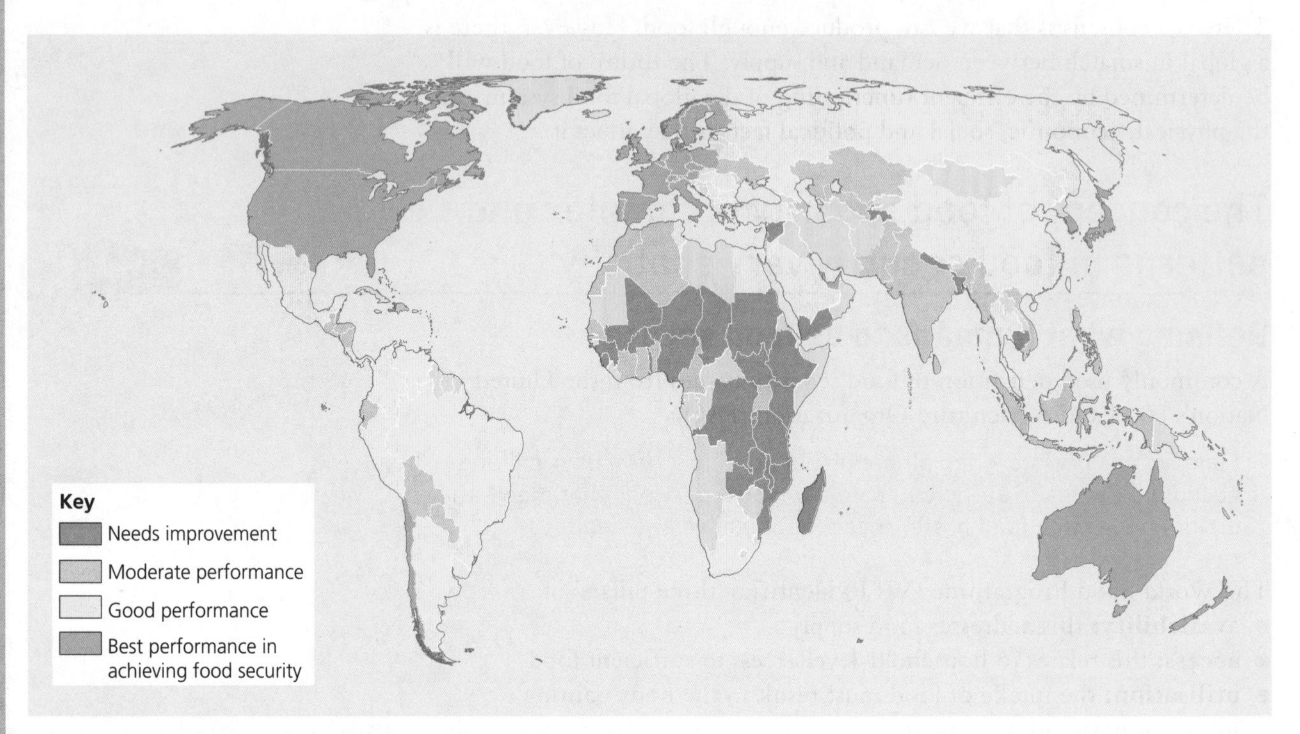

Figure 4.27 Global Food Security Index, 2015

Exam tip

Be aware of different measures and representations when analysing trends and patterns.

Revision activity

Produce a bullet-point summary of the global pattern of food security from Figure 4.27. Organise your notes into three sections:

1. regions and countries with the best performance
2. those with good/moderate performance
3. those needing improvement

Be geographically specific, e.g. the continent of Africa: moderate and good performance in north Africa, e.g. Egypt, but improvement needed in most of central and southern Africa.

Exam tip

It is important that you have developed the skills to describe the pattern of food security shown by a range of data presentation methods. Your data should also be based on current trends.

Variations in food security between and within countries

It is important to remember that food security varies at all scales: between countries and also within countries (Table 4.18).

Table 4.18 Food security variations at the global, national and regional scale

Global food security	National food security	Regional food security within countries
The global regions of Latin America and the Caribbean, Eastern Europe, East and Southeast Asia and North Africa have all made good progress in achieving food security (Global Hunger Index and Global Food Security Index, 2015) North America, Western Europe and Australasia are the best-performing global regions	Global Hunger Index Scores of the best- and worst-performing countries achieving a reduction of their GHI score (1990–2013): ● best performance: Kuwait, Thailand, Vietnam, Ghana and Mexico ● worst performance: Swaziland, Iraq, Burundi, Eritrea and Sudan	Variations in food security within countries happen across the development continuum Examples include: ● food insecurity in eastern provinces of China ● food-insecure groups in urban areas in many countries, e.g. amongst the urban poor in Accra, Ghana and inner-city districts such as the Bronx in New York

Revision activity

Make brief bullet-point notes on variations in food security *within* two contrasting countries.

Food is a precious resource and global food production can be viewed as an interconnected system

REVISED

The physical conditions required for growing food

Physical factors exert a major influence on farming methods despite technological advances. The physical conditions required for growing food include air, climate, soil and water.

Air

Photosynthesis involves the absorption of CO_2 from the atmosphere and the release of O_2. Plants also require some oxygen for respiration to carry out their functions of water and nutrient uptake. Some plants fix nitrogen from the atmosphere.

Climate and soil

The two most important determinants of food production methods are climate and soils. Climate is a major factor affecting soil characteristics and, in turn, the natural vegetation (and food growth) of an area. The main climatic factors affecting food growth are temperature, sunlight and precipitation (Table 4.19). Figure 4.28 shows the location of the world's main farming types.

Exam tip

Make sure you have a clear understanding of how climatic factors affect the type and location of agriculture.

Table 4.19 Climatic factors affecting food growth

Temperature	Sunlight	Precipitation
This is a key climatic factor as each crop type requires a minimum growing temperature and a minimum growing season, e.g. 6°C Several frost-free days are also required for crop growth Within the tropics there is a continuous growing season Temperature and length of growing season both decrease with height above sea level	The process of photosynthesis requires sunlight and, as with temperature, crops vary in their light requirements Both light intensity and duration of sunlight are important for crop growth	The average rainfall in an area determines what crops are most suitable (tree crops, cereals) and whether irrigation is required The effectiveness of precipitation must also be considered, as in very warm climates the rainfall may quickly evaporate In addition to average quantity, the seasonal distribution and type of rainfall is also important for growing crops, e.g. high levels of summer rainfall mean that water is available during the growing season and prolonged periods of rain allow more infiltration than heavy downpours where the water can quickly be lost by run-off and may also cause soil erosion

Revision activity

Soil has a major influence on farming. Make a revision summary on how farming is influenced by the depth, drainage, texture, structure, pH and mineral content of soils.

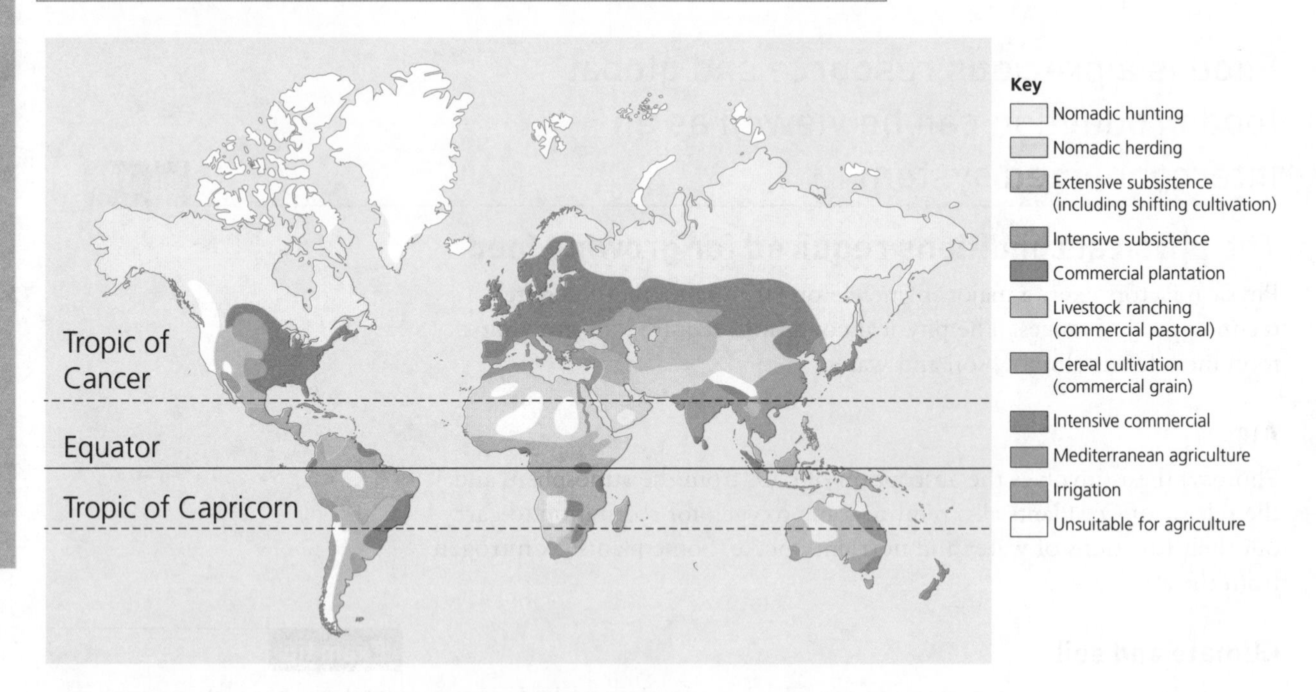

Figure 4.28 Location of the world's main farming types

Typical mistake

When asked to **analyse** a pattern do not just describe it — you have to identify characteristics and give evidence.

Water

Water comprises 80% of living plants and is a major determinant of crop productivity and quality. Water is essential for the germination of seeds and crop growth. In terms of biological functions, water is used in photosynthesis to produce sugars from light energy, and water acts as a solvent and means of transport for minerals and sugars through the plant.

Now test yourself TESTED ☐

46 How does soil contribute to plant growth?
47 How can evaporation affect plant growth?

Answers on p. 223

Feeding the world is a complex system

Agricultural systems in different parts of the world are increasingly interconnected to form what is referred to as the **global food system**. The global food system is a complex network which includes the production, harvest, processing, transport and consumption of food and the disposal of waste. A key issue is that by buying more food than we eat, land and resources are used that could be used to feed people suffering food insecurity. 'Dealing with' food waste should focus on more effective management and *distribution* of food.

The global system does not function as an efficient 'whole' — there are areas of food surplus and food scarcity which have emerged across the development continuum.

The system is influenced by the policies of individual nation states; by trade; by the actions of organisations such as FAO, the World Food Programme and the World Bank; and by a range of **MNC**s.

An **MNC** (multinational corporation), like a TNC, is a large company with operations in several countries.

Typical mistake

Waste is not just a problem in ACs. Waste can exist at the point of production and consumption. In LIDCs food waste can result from crop disease and inefficient storage.

How food production methods vary from intensive to extensive and subsistence to commercial

Food is produced through a range of farming systems which are classified into groups. The criteria by which farming systems are classified include environment, location, tenure, inputs, outputs and market. A range of food production methods therefore exists (Table 4.20).

Typical mistake

Subsistence farming often includes an element of some crops being sold for cash to buy essentials such as clothing. Often, students assume that all crops are consumed by the producer.

Table 4.20 Food production methods

Arable and pastoral	Subsistence and commercial	Shifting and sedentary	Extensive and intensive
Arable: the growing of food crops, often on fairly level, well-drained soils, of good quality **Pastoral:** the raising of livestock, often in areas unsuitable for arable farming (too cool, wet or dry or too steep). Soils often have limited fertility. Livestock farming is sustainable only when the **carrying capacity** of the area is not exceeded	**Subsistence:** provision of food by farmers for their own consumption and for the local community. Subsistence farmers are vulnerable to food shortages because of the lack of capital and other entitlements **Commercial:** farming for profit, often on a large scale with high capital inputs	**Shifting cultivation:** confined to a few isolated places with low population density, large areas of land and limited food demands (e.g. indigenous groups in tropical rainforests). The system, which is sustainable at low population densities, is essentially a rotation of fields rather than a rotation of crops **Sedentary:** farmers remain in one place and cultivate the same land year after year	**Extensive:** large-scale commercial farming; inputs of labour and capital are small in relation to the area farmed. Yields per hectare are low but yields per capita are high **Intensive:** small-scale with high labour and/or capital inputs and high yields per hectare
Examples: Arable farming: the Nile Valley, the Great Plains Pastoral farming: hill sheep farming in Wales, nomadic herding in the Sahel	Examples: Subsistence: wet-rice farming in India Commercial: cattle ranching in South America, oil palm plantations in Malaysia	Examples: Shifting cultivation: the Amazon Basin and Indo-Malayan rainforest Sedentary: dairy and arable farming in the UK	Examples: Extensive farming: Canadian Prairies cereal farming Intensive: horticulture in the Netherlands

Globalisation is changing the food industry

REVISED

The influence of globalisation on the food industry

- Food distribution and consumption have become increasingly global despite the fact that production is a local process determined by factors such as climate, soil and growing season.
- Developments in transport, communications and refrigeration technology have enabled the development of global food chains. Perishable food products can now be transported over long distances.
- Complex transnational food production networks now exist between producers and consumers.
- Food demand has changed as incomes rise through economic growth and populations become more urbanised.
- Food tastes are changing across the world, especially in the affluent markets in North America, Europe and parts of Asia.
- A wide variety of foods are now available from across the globe, with all-year-round provision.

Now test yourself

TESTED

48 What factors have led to the globalisation of the food industry?

Answers on p. 223

Globalisation of the food industry creates a number of issues and opportunities

Issues	Opportunities
• **Food miles** Food miles are a measure of how far food has travelled from producer to consumer. Improved transportation and technology mean that food now travels over long distances from producer to consumer. This is often by air transport, which raises environmental concerns over greenhouse gas emissions • **Inequality between TNCs and small suppliers** The production, distribution and consumption of food have become part of a global industry dominated by transnational companies, **agribusinesses** and large food retailers. TNCs often favour large, capital-intensive operations leaving small producers disadvantaged and marginalised. Also the scale of the involvement by TNCs makes it difficult for national governments to regulate their own food systems • **Obesity** As countries develop and affluence increases, consumption of food shifts from cereals to more expensive foods such as meat and dairy products. Development and urbanisation also lead to an increase in fast-food outlets. Weight gain and obesity can result. The rising middle classes in EDCs such as China and Brazil have been significantly affected • **Price crises** Global food prices are volatile and vulnerable to **price shocks**. Price increases may be due to a sudden shortage in supply, transport issues, fuel cost or natural hazards	• **Technological innovation** There have been significant technological advances in food production, but it remains a challenge to ensure that these advances are shared carefully between all farmers and between countries at varying stages of development • **Short-term food relief** Globalisation has facilitated the provision of food aid. Types of international food aid include bilateral (between two countries), multilateral (provided by a number of countries and agencies) and non-government aid • **Consumer choice** Global food trade has meant that many consumers now have access to a wide range of products through retail outlets and online. Many food brands have global appeal, such as Coca-Cola. Retail giants such as Carrefour have extended consumer choice through multinational locations

Figure 4.29 Comparison of the issues and opportunities created by the globalisation of the food industry

Agribusiness is a large-scale farming practice, run on business lines.

Price shock: an unexpected and unpredictable change in prices which can be positive or negative.

Revision activity

Make a list of the advantages and disadvantages of (a) technological innovation in food production and (b) short-term food relief.

Now test yourself

TESTED

49 Why has obesity become a global issue?

Answers on p. 223

Exam tip

Remember that the issues created by the globalisation of the food industry impact on countries across the development continuum.

Exam tip

You should be able to refer to named examples for all the issues and opportunities outlined in Figure 4.29, e.g. genetically modified crops as an example of technological innovation.

What are the causes of inequality in global food security?

A number of interrelated factors can influence food security

REVISED

Physical factors that affect food security

There are a range of environmental factors that affect food production and therefore the physical supply of sufficient quantities of food. These include geology, soil, length of growing season, temperature, precipitation and water supply, sunlight, altitude, aspect and slope.

Some physical factors are covered in the section 'The physical conditions required for growing food' on p. 180. Others include altitude, aspect and slope (Table 4.21).

Typical mistake

Do not view these physical factors in isolation — remember that they are interlinked, e.g. geology, climate and slope all affect soils.

Table 4.21 Additional physical factors affecting food security

Altitude	Aspect	Slope
Altitude is a good example of how climate, soils and growing season are all interlinked. As height increases, temperature decreases, snow and precipitation increase and the growing season decreases. At the same time, soils take longer to develop, nutrient recycling is slower and leaching becomes prevalent	Mountains are characterised by steep slopes with different directions of aspect. Aspect determines **microclimate**: south-facing slopes receive more sunlight than north-facing slopes and are therefore warmer with drier soils. Crops on south-facing slopes can grow at higher altitudes than those on north-facing slopes. Evapotranspiration rates and temperatures are lower on north-facing slopes	Slope angle affects rates of soil erosion, use of machinery and soil formation (depth and drainage). On steep slopes soils are often thin, poorly developed and excessively drained. Soils at the base of a slope can become waterlogged. On gentler slopes, there is less erosion and **leaching**

A **microclimate** is a local climate determined by topography and land use.

Leaching occurs when soluble materials drain away in a soil.

Now test yourself

TESTED

50 How does slope affect soil characteristics?

Answers on p. 223

The social, economic and political factors affecting food security

It is the influence of humans and their effective use of the physical resources that will determine how successful food production is and the extent to which food security can be reached. The social, economic and political factors affecting food security are outlined in Figure 4.30; as with physical factors there is overlap between categories.

Exam tip

You should have a general understanding of the global distribution of the main farming types: see Figure 4.28, p. 181.

Economic factors	Political factors	Social factors
Competition Competition exists in a range of contexts within the food supply chain. Two of the main forms of competition are: • competition in food markets: the growing dominance of agribusinesses and TNCs in global food supply has reduced competition • competition for scarce resources: there is increasing competition across the globe for essential agricultural resources such as land, water and energy **Farm size** Different economic factors can affect farm size. In the developed world, the growth of agribusinesses and the use of farm machinery has led to increasing farm size. In some parts of the world as population increases farms are being subdivided amongst growing families **Transport** With developments in transport networks, food products now travel long distances over short time spans. However, this does have environmental consequences regarding the increasing air miles of some fresh food produce **Markets** Market demand relies on a range of factors such as affluence and taste. With advances in transport it is less important today for farms to be close to markets. However, this will be a factor in less developed parts of the world where infrastructure is not as developed and storage is limited **Capital** Farming systems can be capital-intensive, requiring large investments of both finance and machinery, or they can be labour-intensive, with less reliance on capital. Farmers in some developing countries lack the support of financial institutions and borrowing money for investment in the farm is either very difficult or very costly **Technology** Technological advancements can range from new strains of seeds and fertilisers to advances in mechanisation and irrigation techniques. Capital reserves are needed to take advantage of technological advancements and so sometimes they can leave poorer farmers and LIDCs at a disadvantage	**Land ownership systems** This refers to the ownership rights that farmers have on their land. There are many different types of land ownership and they vary considerably across different countries according to political and economic factors. Types include: owner-occupiers, tenants or landless labourers with no ownership rights. Food security is impacted by land ownership as it can affect productivity and decision making, and also the distribution of the harvest will usually be determined by the land owner. **Land grabbing** Land grabbing is a process whereby rich countries acquire land in poorer countries: investor countries and target countries. Investor countries may either have land and water constraints but high levels of capital, e.g. Saudi Arabia, or they may have specific food security issues such as very large populations, e.g. China **Government policy** Government intervention in the farming system can range from a high level of intervention in a centrally planned economy to a situation where governments struggle to intervene because of the influence of powerful agribusinesses and TNCs. Sometimes intervention is through a multinational agreement such as the EU	**Inheritance laws** Cultural factors in some countries mean that farms are divided amongst siblings as inheritance or as dowry customs. This may subdivide farms

Figure 4.30 The economic, social and political factors affecting food security

Revision activity

Use the comparison diagram (Figure 4.31) to evaluate the costs and benefits of land grabbing for the target countries, e.g. an increase in agricultural employment but unequal power relations.

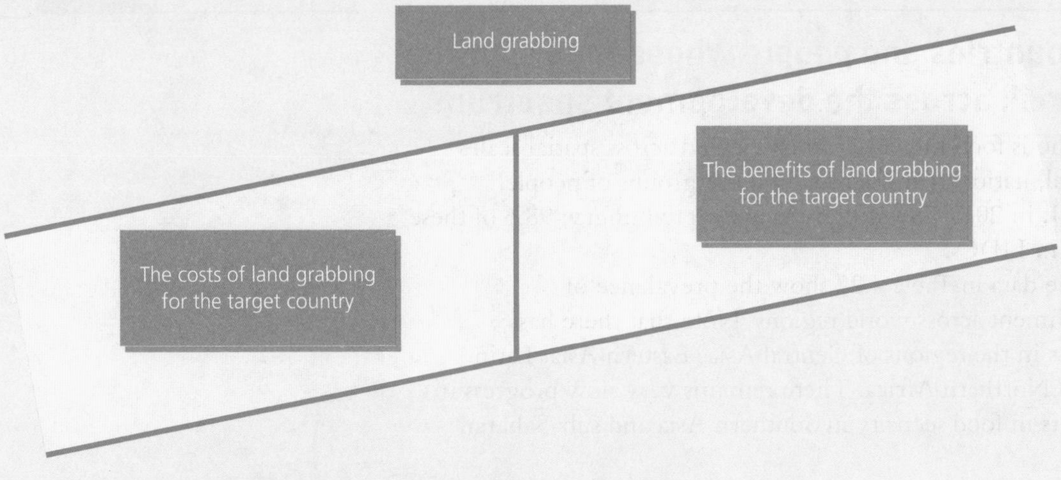

Figure 4.31 Comparison revision diagram (after David Didau)

Theoretical positions on food security

Thomas Malthus (1798)

Malthus' theory is that an optimum population exists in relation to food supply and that an increase beyond this will lead to 'war, famine and disease'. Malthus stated that, in the absence of checks (e.g. famine and war), human population will grow at a geometric rate: 1, 2, 4, 8, 16 and so on, doubling every 25 years. Food supply, at best, can only increase at an arithmetic rate: 1, 2, 3, 4 and so on, and is therefore a check on population growth. Given a limit to the amount of food that a country can produce, Malthus suggested preventative measures to limit population growth, e.g. abstinence from marriage.

Since Malthus' theory was put forward, food production has been increased by:

- high yield variety (HYV) crops
- new foods such as soya
- use of agrochemicals
- land acquisition, e.g. drainage of wetlands

Esther Boserup (1965)

Boserup believed that countries have the resources, knowledge and technology to increase food supply in response to growth in population and that population growth is needed to trigger such advancements.

Exam tip

Land grabbing affects food security for *both* the target and investor countries. This point should be illustrated by reference to examples.

Exam tip

Remember that the theories of Malthus have gained recent support with demographers, known as neo-Malthusians, based on recent evidence of famines, wars and water security.

Typical mistake

Present a balanced view of theoretical positions — they do have some relevance today and always form a useful starting point for discussion.

Case study of **one** place to illustrate how human and physical factors are/have combined to cause issues with food security

Online (see p. 3) you will find a case study on India to illustrate the above.

What are the threats to global food security?

Risks to food security

REVISED

Regions, countries and people whose food security is most at risk across the development spectrum

The issue of who is food insecure can be viewed across spatial scales (global, regional, national) and across different groups of people.

- **Global level:** in 2015, 795 million people were hungry; 98% of these people lived in LIDCs.
- **Regions:** the data in Table 4.22 show the prevalence of undernourishment across world regions. Note that there has been progress in the regions of Central Asia, Eastern Asia, Latin America and Northern Africa. There remains very slow progress in improvements in food security in Southern Asia and sub-Saharan Africa.

- **Countries:** food security remains fragile in some countries despite good progress in recent times, e.g. Algeria, Egypt, Morocco and Tunisia in northern Africa. In other countries, natural and human disasters and political instability have contributed to crisis situations, e.g. Syria, Yemen, Nepal and Iraq. Table 4.18, p. 180, column 2, shows the best- and worst-performing countries between 1990 and 2013 as measured by the Global Hunger Index.
- **People:** distinct groups are vulnerable to food insecurity, including:
  - rural dwellers: 75% of hungry people live in rural areas, mainly in Africa and Asia, and are mostly people dependent on agriculture with no other source of income
  - farmers: half of the world's hungry people are from small-scale farming communities farming on marginal land
  - children: an estimated 146 million children in LIDCs suffer from acute or chronic hunger
  - women: evidence shows that women are more affected by hunger than men, which has further impacts on infant mortality and low birth-weight children

Table 4.22 Number of undernourished (millions) and prevalence (%) of undernourishment around the world (Source: FAO)

Region	1990–1992		2014–2016*	
	No. (millions)	%	No. (millions)	%
World	1,010.6	18.6	794.6	10.9
Developed regions	20.0	< 5.0	14.7	< 5.0
Developing regions	990.7	23.3	779.9	12.9
Africa	181.7	27.6	232.5	20.0
Asia	741.9	23.6	511.7	12.1
Latin America and the Caribbean	66.1	14.7	34.3	5.5
Oceania	1.0	15.7	1.4	14.2

*2016 estimate.

Revision activity

Make bullet-point notes or annotate a copy of Table 4.22 to summarise the regional pattern of undernourishment.

Exam tip

The global pattern of food insecurity is complex and ever-changing, so make sure that you have up-to-date examples of regions and countries where food security is at risk.

Revision activity

Select two countries from your studies and make bullet-point notes giving some data reflecting food insecurity and the reasons for the situation. Try to pick countries that will give contrast: e.g. Syria has suffered as a result of civil unrest, while Nepal is vulnerable because of natural hazard events.

The issue of geographical pinch points

The **food supply chain** is the means by which food is transferred from the farm to our plates. It is a complex flow of food produce between sites of production, processing, distribution and consumption. **Pinch points** are places in the chain where disruption occurs. Reasons can be political, economic, social, environmental or technological and they can occur at the global, regional and local scale.

Revision activity

On a copy of a blank world base map, annotate some examples of pinch points.

Food supply chain: the process whereby food moves from producer to consumer. It involves production, processing, distribution, storage, consumption and disposal.

A **pinch point** is a point where congestion is likely to occur.

Desertification occurs when there is a reduction in agricultural productivity due to overexploitation of resources and natural processes such as drought.

The physical and human causes of desertification and how this changes ecosystems to increase risk to food security

Desertification now ranks amongst the greatest environmental challenges of our time, with 168 countries worldwide and in total 15 billion people affected.

The situation is acute in countries such as Somalia, Ethiopia, Djibouti and Kenya, where the combination of lack of government focus and prolonged periods of drought linked to climate change is driving desertification levels. Table 4.23 summarises the causes of desertification.

Table 4.23 The causes of desertification

Human causes of desertification	Physical causes of desertification
Poverty: lack of farming investment and lack of money force people to farm any available land, pushing them onto marginal land such as steep slopes; the land is also often overused	**Climate change:** increased periodic drought and changing rainfall patterns damage animals' habitats and soil quality. Crops die and farming practices change; land degradation intensifies
Changing farming practices including overgrazing and expansion of cropped areas: new more intensive agricultural systems deplete soil nutrients as fallow periods and crop rotation are abandoned. On marginal land soils are often already fragile	**Soil erosion by wind and water:** this removes the top layer of soil which contains nutrients
High demand for irrigation water: irrigation is a human response to water shortages. Water scarcity causes crops to die and poor farming practices to be used. Irrigation projects are also poorly managed, unsuitable and underfunded	**Salinisation:** this is a process involving a combination of physical and human factors (see p. 192). It is a problem because few plants are salt tolerant
Demand for fuelwood: as land is cleared for fuelwood the soil is left exposed to wind and water erosion	
Political and economic instability: political instability sometimes leads people to stay in the same areas and land becomes overused; in other cases political and economic instability leads to displacement of people or individuals (male out-migration), widening the problem and, in the case of male out-migration, leaving women to undertake intensive farming close to the home	

Impact of desertification on ecosystems

- Reduction in vegetation leads to reduced habitats and increased competition.
- Continual cropping decreases nutrient recycling.
- Soil nutrients are lost through wind and water erosion.
- Carbon sinks are reduced.
- There is loss of biodiversity.
- Food webs become fragile.

Now test yourself

TESTED

51 What is desertification?
52 Explain how population growth can lead to desertification.
53 State **three** impacts of desertification on ecosystems.
54 Draw a flow diagram to explain how desertification is linked to poverty.

Answers on p. 223

Exam tip

Remember that famine rarely means an absolute food shortage for everyone. It often impacts the poorest section of society with the lowest level of food security.

Case study of one dryland area

Revision activity

Make your own case study summary of a dryland area covered in class. Focus on:

- how food security is influenced by the specific ecosystem, climate and hydrology
- worsening factors such as population change, land grabbing and climate change

The food system is vulnerable to shocks that can impact food security

REVISED

How climate change is leading to increasing frequency of extreme weather events which can affect food production

Extreme weather events include heatwaves, droughts, wildfires, periods of intense rainfall, floods, hurricanes and tornadoes.

The increased frequency of these is linked to climate change because: warming of the atmosphere increases the number of times that temperatures reach extreme levels, more water vapour evaporates from the oceans → an increase in water vapour in the atmosphere → more intense rainfall → the retention of heat energy from the sun → further global warming.

Revision activity

Make summary revision notes on examples of how extreme weather events such as heatwaves, drought, floods and tropical storms can affect food production.

Exam tip

Always give a balanced response to climate change questions regarding the impact on agricultural production, as some areas will have higher yields because of a warming climate.

How water scarcity can affect food production

Agriculture accounts for 68% of water drawn from rivers, lakes and aquifers. Of the water available, up to 60% is lost through poor irrigation systems and high levels of evapotranspiration. Water scarcity causes crops to die and poor farming and irrigation practices to be used. Many countries across the development spectrum are trying to find more efficient means of using water in agriculture. One such measure relates to the transfer of **virtual water**. In water-scarce countries, importing water-intensive food products, e.g. lettuce, relieves pressure on domestic water resources. Another measure includes countries offering farmland to water-scarce countries but this can compromise food security in the host country.

Virtual water is the volume of fresh water needed to make a product, measured at the place where the product was manufactured.

How tectonic hazards can influence food production and distribution

Tectonic hazards can have a positive impact on food production such as the high levels of fertility of volcanic soils. However, in terms of shocks that can impact food security there are negative impacts on both production and distribution. These are summarised in Table 4.24.

Table 4.24 How volcanic eruptions and earthquakes can affect food security

Volcanoes	Earthquakes
Volcanic ash-fall destroys pasture land Volcanic ash can increase the levels of sulphur in soils and lower pH to levels where plants cannot survive Ash-fall can adhere to fruit skins and either destroy crops or mean that it is too expensive to clean them for sale to foreign markets with rigid standards	Transport and food distribution may be disrupted by cracks from earthquake activity Food stocks may be destroyed Livestock may be killed There may be damage to irrigation systems

Now test yourself

TESTED

55 How does the increased frequency of heatwaves affect climate change?

56 What is virtual water?

Answers on p. 223

Case study of **one** indigenous farming technique in an extreme environment

Online (see p. 3) you will find a case study of the Arctic to illustrate the above.

How do food production and security issues impact people and the physical environment?

Imbalance in the global food system has physical and human impacts

REVISED

Agriculture is one of the world's largest users of land, but this has come at a cost to the environment. Global food supply has increased over recent years as a result of higher yields per unit of land; crop intensification; an increase in the amount of land being farmed; and technological advances — all of which have impacts on both the physical environment and humans. A summary of impacts of food production on the physical environment is shown in Figure 4.32.

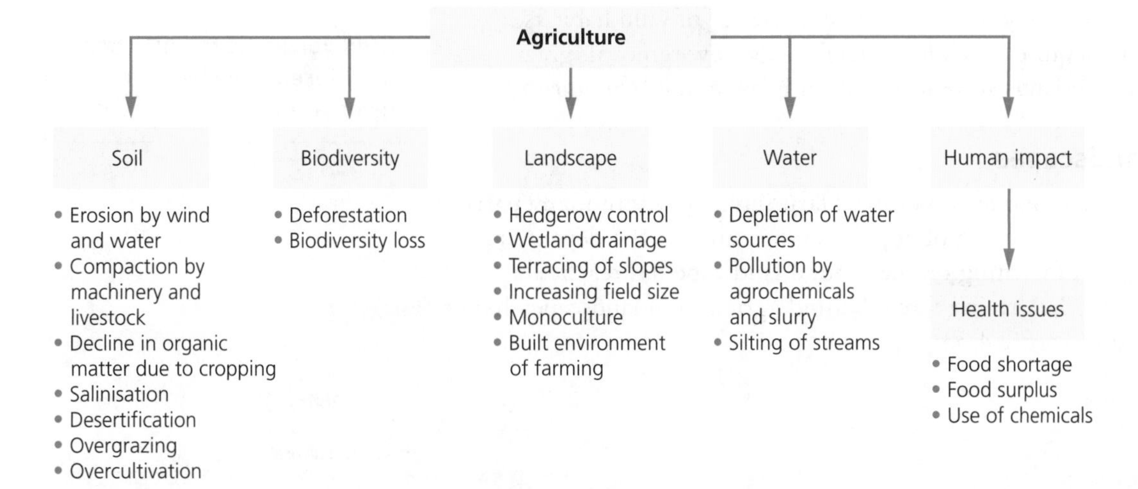

Figure 4.32 Summary of the environmental impacts of food production

How attempts to increase food production and security can impact the physical environment

Irrigation and salinisation

Salinisation is a natural process often made worse by human activity. The process can be explained as follows:

Low precipitation and high evaporation → salts in the soil are brought to the surface → plants intake water but leave salts behind → a salt layer is left which is toxic to plants and can inhibit water absorption and directly affect plant growth → land is consequently unusable for agriculture.

Poor irrigation techniques can intensify salinisation when water is brought to land that is naturally dry because:

- irrigation can lead to increased rates of groundwater recharge causing water tables to rise; this brings salts to the plant root zone which affects both plant growth and soil structure
- water used to irrigate crops evaporates quickly in very dry conditions
- inefficient irrigation with high levels of leakage raises water tables and increases the risk of salinity
- excessive withdrawal of underground water in coastal areas leads to infiltration of saline water into fresh groundwater supplies

Impacts of irrigation salinity include:

- reduced agricultural productivity
- reduced income for farmers
- decline in water quality with a potential impact on livestock and future irrigation
- soils with weakened structure, nutrient loss and increased erosion
- decline in natural vegetation with a consequent loss of habitat for some species

Salinisation is the accumulation of salts in soil.

Revision activity

Draw a simple diagram to explain the process of salinisation.

Exam tip

Diagrams can be used very effectively in exam answers. They can present a clear and concise explanation, saving a lengthy block of text. If used, make sure that diagrams are clear and well annotated; numbering the sequence of an explanation also helps.

Deforestation and the impacts on biodiversity

Increasingly, natural forest is being converted into agricultural land. In the tropics this has accounted for 80% of new agricultural land. The increase of large agribusinesses taking over from small farmers has meant that the practice is now on a much larger scale. LIDCs without valuable resources such as oil and gas depend on agriculture to support the economy. As a result, conversion of forest to agricultural land continues at a rapid pace.

Species richness is closely related to the occurrence of wild habitats. Deforestation to acquire agricultural land, hedgerow removal, grazing and drainage of wetlands all reduce natural habitats and **biodiversity**.

Biodiversity is the number of different species in a given area.

Changing landscapes

Because of the rise of modern industrial farming methods in many parts of the world and also because of population pressures in the developing world, the impact of farming on the natural landscape has reached a concerning level. The driving forces behind agricultural landscape change are summarised in Figure 4.33.

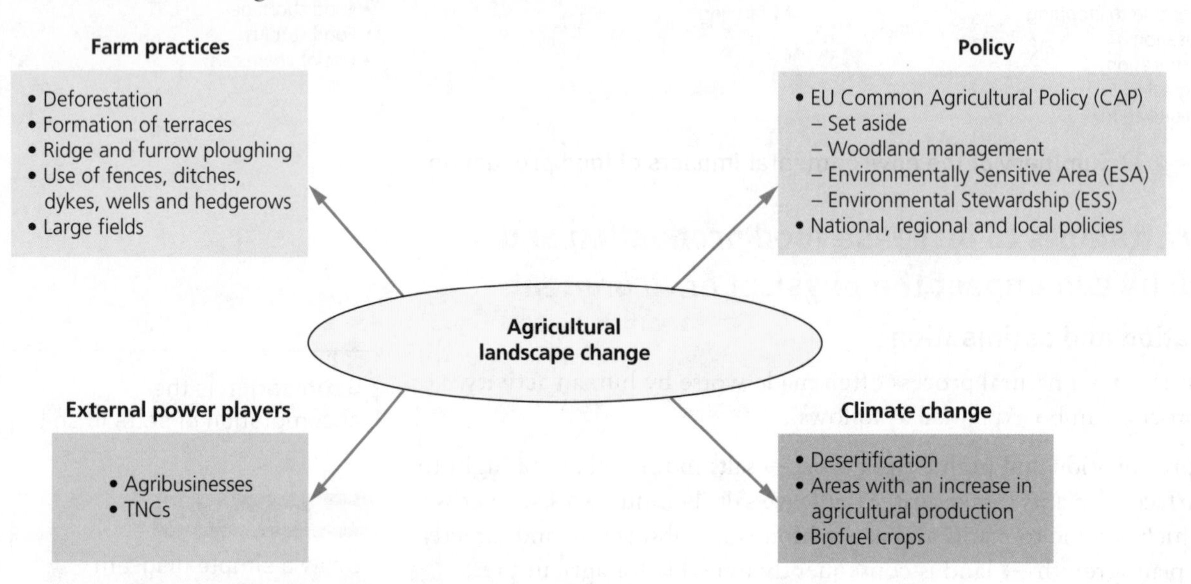

Figure 4.33 The driving forces in agricultural landscape change

Water quality and agrochemicals

Agrochemicals such as herbicides, insecticides and pesticides have become increasingly important in crop production. Intensive use of these chemicals leads to environmental problems, such as contamination of soil and groundwater. When applied, only around 15% of the preparation hits the target; the rest is distributed in the soil and air — pesticide drift. Emerging economies are having particular problems as there is less monitoring and control. Impacts of agriculture on water quality include:

- **eutrophication**
- leaching of nitrates into groundwater
- contamination of aquifers
- contamination of surface water and biota from the run-off of pesticides

Agrochemical: a chemical used in agriculture, such as pesticides and fertilisers.

Eutrophication is the process whereby nutrient enrichment in water (streams, rivers, lakes) leads to a fall in oxygen levels and the subsequent death of species which are dependent on oxygen.

Now test yourself

TESTED

57 How does the addition of fertiliser to agricultural land alter freshwater ecosystems?

Answers on p. 223

Case study of how one physical environment is/has been impacted by food production methods

Revision activity

Make revision notes on a case study covered in class for this part of the course. Remember to focus on the specific short- and long-term impacts on the environment.

How food security issues impact people

Health issues associated with food shortages

Food security refers not only to access to sufficient food but also to safe and nutritious food that meets dietary needs and helps maintain a healthy life. Nearly 30% of the world's population suffer from **malnutrition** when daily calorie intake is low, leading to health issues. However, increasingly, a different set of health issues are associated with excessive calorie intake. The health issues associated with high and low food consumption are set out on the 'nutrition spectrum'.

Malnutrition occurs when an unbalanced diet results in shortages of proteins and essential vitamins.

Exam tip

Make sure that you understand the different health impacts of being under- and overweight.

Health issues associated with food surpluses and poor diet

The number of overweight people now exceeds those underweight. Obesity affects several developing nations, e.g. Colombia has an obesity rate of 41% and Brazil 36%. Such is the spread that the term **globesity** is now used.

Globesity is the global increase in higher than average weight resulting from poor diet and lack of exercise.

The cause of obesity is an energy imbalance between calories consumed and calories expended. Also, there has been increased intake of energy-dense foods high in fat and sugars and low in vitamins and minerals, e.g. processed foods, ready meals and fast foods.

Health consequences of obesity include non-communicable disease such as diabetes, cardiovascular diseases and some cancers. Children with poor diet and a lack of exercise who become overweight risk health consequences such as fractures and early markers of cardiovascular disease.

Harmful impacts on human health as a result of the increased use of chemicals and pesticides

Globally the use of insecticides, pesticides and chemicals has increased to control harmful crop diseases that threaten yields and to increase overall food production. Monitoring and management of the impacts of **agrochemicals** has led to a range of health scares.

In ACs such as the UK, organisations including the Food Standards Agency (FSA) exist to protect public health regarding food safety issues. In EDCs and LIDCs there is less regulation, which is of growing concern.

Case studies of two places at contrasting levels of economic development to illustrate the implications of poor food security for the lives of people

Revision activity

Make your own revision notes on the two places covered in class which illustrate how poor food security has implications for people's lives, e.g. their health, finance, fitness or ability to work and learn.

Is there hope for the future of food?

Food is a geopolitical commodity; a number of key players will continue to influence the global food system

REVISED

The geopolitics of food

A series of events such as the global economic recession, food supply shocks, civil unrest, food riots and concern over the long-term food supply as a result of global warming have meant that food supply is increasingly affecting and being influenced by political decisions and events. **Geopolitics** refers to ways in which geographical factors shape international politics.

Key players in the global food system include:

- national governments
- international organisations such as the World Trade Organization
- profit-making organisations: agribusinesses, TNCs and food retailers
- non-governmental organisations (NGOs) such as the World Fair Trade Organization

Exam tip

The global food problem is not caused by scarcity of food — it is a matter of the distribution of food resources and access to markets, technology and basic resources such as land and water, which are increasingly affected by politics.

Now test yourself

TESTED

58 Explain how obesity can affect countries at different stages of the development continuum.

59 Give **two** examples of how food has become a geopolitical commodity.

Answers on p. 223

The opportunities between countries to ensure food security

Agricultural trading policies

Trade in food is needed to ensure global food security. Agriculture accounts for more than one-third of export earnings in 50 developing countries. However, export subsidies and import tariffs by developed countries mean that some poor countries are unable to compete in international markets.

Different types of trade agreement exist. They are summarised in Table 4.25.

Table 4.25 Different types of trade agreement

Trading bloc	Multinational agreement	Bilateral trade agreement
An agreement between a number of countries to promote free trade among its members	Several countries engage in a trading relationship with a third party	A trade agreement between two political entities that has mutual benefits and is legally binding
Tariffs are imposed on the imports from non-member states Example: the EU	Example: African, Caribbean and Pacific Group of States (ACP) nations being given free trade access to EU markets	Example: Sainsbury's trade agreement with St Lucia for fair trade bananas

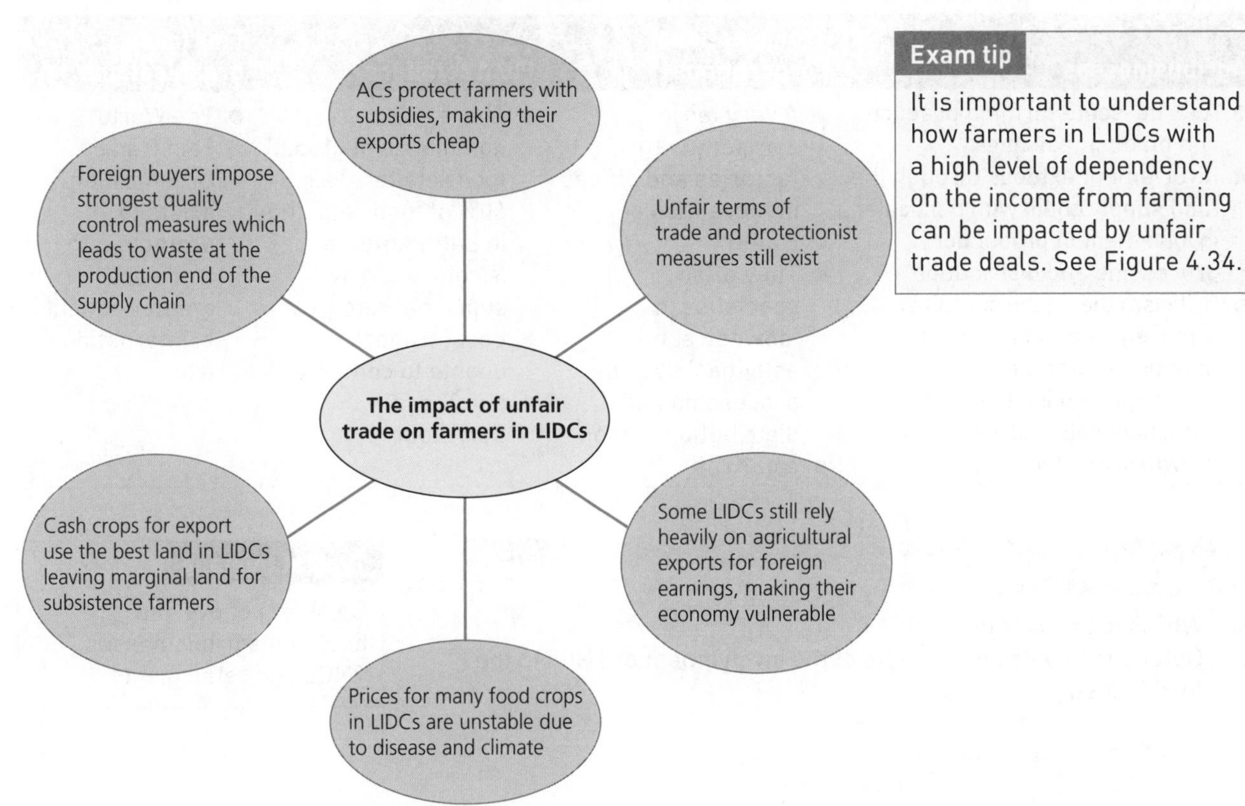

Figure 4.34 The impact of unfair trade on farmers in LIDCs

Exam tip

It is important to understand how farmers in LIDCs with a high level of dependency on the income from farming can be impacted by unfair trade deals. See Figure 4.34.

The role of the World Trade Organization (WTO)

The main role of the WTO is to provide a forum for governments to negotiate trade agreements and to settle trade disputes. Criticisms include not doing enough to achieve the equitable distribution of food.

Appropriate aid

Types of aid include:

- project food aid
- programme food aid
- emergency or relief food aid

Criticisms of food aid include the following:

- Donor-driven food aid centres on the use of food aid as a means of 'dumping' surplus food from ACs.
- Food aid dependency could be a long-term outcome of food aid for the recipient countries. Large quantities of food aid can swamp local markets and drive down prices, reducing the income of indigenous farmers.

Food aid has benefits particularly in crisis situations, e.g. in the civil conflict in Syria and the earthquake in Nepal (2015).

The alternatives include appropriate technology and small-scale projects to promote the recovery of farming. Examples include the MERET programme in Ethiopia which helps farmers reclaim **degraded land** using simple techniques such as terracing hillsides to prevent soil erosion.

Degraded land is land where the production capacity has been significantly reduced.

The role and responsibilities of key players in influencing the global food system

Table 4.26 Key players influencing the global food system

Agribusinesses	Trans-national corporations	Food retailers	Fair trade organisations
• A large-scale farming practice run on business lines • Involvement extends through the food supply chain with particular involvement in production, processing and distribution • Criticism that in the pursuit of profit, environmental issues have been compromised, especially through their use of agrochemicals and hormone growth promoters	• A very large company with factories and offices in more than one country • They often specialise in 'downstream activities' such as processing and distribution of food • e.g. Kraft	• There is a dominance of global food retailers, e.g. 60% of food retailing in Latin America is controlled by supermarkets • Local traders are unable to compete • e.g. Tesco, Carrefour	• The World Fair Trade Organization exists to promote fair trade and greater equality in international trade

Now test yourself

TESTED

60 Why is food aid criticised?
61 Outline the costs and benefits of the involvement of TNCs in the food industry.

Answers on p. 223

Typical mistake

Be aware of overlap between agribusinesses, TNCs and retail giants.

There is a spectrum of strategies that exist to ensure and improve food security

REVISED

Approaches to increasing food security can vary

A complex range of physical and human factors threaten food security, e.g. climate change, natural hazards, land degradation, financial crises, unfair trade deals, competition and price rises. As a result, a range of approaches also exist to ensure long-term food security.

Short-term relief

This approach would include methods such as food aid administered in a crisis situation to alleviate immediate risks to food supply. Long-term dependency on food aid can cause problems.

Capacity building

This refers to the capability of countries and communities to build a resilient food supply system that can withstand threats such as those listed above. It can be achieved by economic development and access to fair trade deals, for example.

Typical mistake

Do not underestimate the ability of indigenous groups to understand how to cope with food security issues. The effective use of local knowledge can be extremely important.

Long-term system redesign

This approach refers to a country's long-term strategic plan to achieve and maintain food security, from securing long-term supply to educating a population in lifestyle choices to ensure a healthy, balanced and nutritious diet.

The effectiveness and sustainability of a range of techniques that exist to improve food security

Examples of the approaches outlined above are summarised in Figure 4.35.

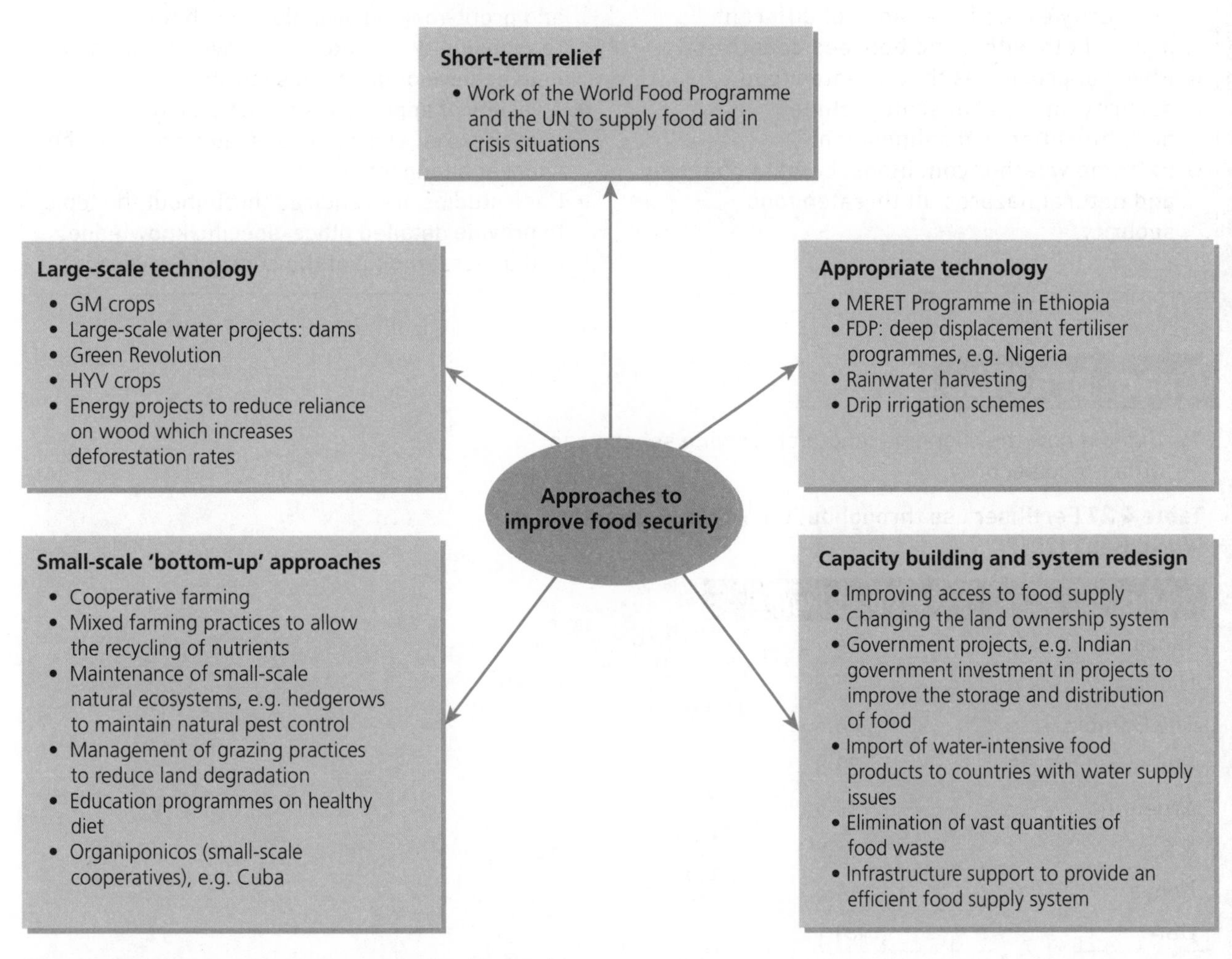

Figure 4.35 Approaches to improve food security

Case study of two contrasting places at different levels of development and the strategies and techniques that have been used to ensure or improve food security

Revision activity

Choose an appropriate method to make summary notes of the two places studied in class to contrast strategies to improve food security. Remember to focus on specific strategies and techniques and back up examples with up-to-date, place-specific facts.

Exam tip

Make sure that you understand the terms used to describe different techniques to improve food security, e.g. appropriate technology, capacity building.

Summary

- Food security is a complex and contested matter with distinctive global patterns.
- Globalisation has had wide-ranging impacts on global food security, which include both opportunities and issues.
- Food security is affected by physical and human factors.
- Despite increases in food production, food insecurity exists for a range of different people both within and between countries.
- Physical processes that threaten food security on a global scale include desertification and salinisation.
- Extreme weather conditions, climate change and natural hazards all threaten food security.
- Impacts of food security on the physical environment include land and soil degradation, biodiversity reduction, changing landscapes and declining water quality.
- People experience a range of health impacts as a result of food insecurity.
- Food has become a geopolitical commodity.
- Opportunities exist to improve food security and profit-making organisations have a responsibility to protect the environment and help achieve global food security.
- A variety of approaches to increasing food security exist at a range of scales and over the short and long term.
- Case studies are required throughout the topic to provide detailed place-specific knowledge and understanding of the concepts covered.

Exam practice

16 Explain how the changing length of growing season can affect food security. [4]

Table 4.27 Fertiliser use throughout the world (kg/ha). World Bank (2012)

Country	Fertiliser (kg/ha)
Tanzania	4.4
Nigeria	5.8
The Gambia	6.5
Ethiopia	23.8
Argentina	38.8
Malawi	39.9
Kenya	44.3
USA	131.1
France	136.9
Chile	358.4
China	647.6
Colombia	744.3
New Zealand	1,485.8
Malaysia	1,570.7

17 Using the data for fertiliser use shown in Table 4.27, calculate the interquartile range (IQR). You must show your working. [4]

18 Using evidence from Table 4.27, analyse the contrasts in the use of fertiliser. [6]

19 Examine how the globalisation of food can impact landscape systems. [8]

20 'Continued technological advances will prevent a Malthusian crisis.' How far do you agree with this statement? [12]

Answers and quick quiz 4D online

ONLINE

Option E Hazardous Earth

A **hazard** is a threat (natural or human) that has the potential to cause loss of life, injury, property damage, socio-economic disruption or environmental degradation. **Natural hazards** lie at the interface between physical and human geography.

What is the evidence for continental drift and plate tectonics?

There is a variety of evidence for the theories of continental drift and plate tectonics

REVISED

Theories of continental drift and plate tectonics

The structure of the Earth

The Earth's cross-section is shown in Figure 4.36. It comprises a sequence of shells, as follows.

- **Core:** the centre of the Earth, an iron-nickel mass that gives the Earth its magnetic field. The inner core is 1,250 km thick. The outer core is liquid and is 2,200 km thick.

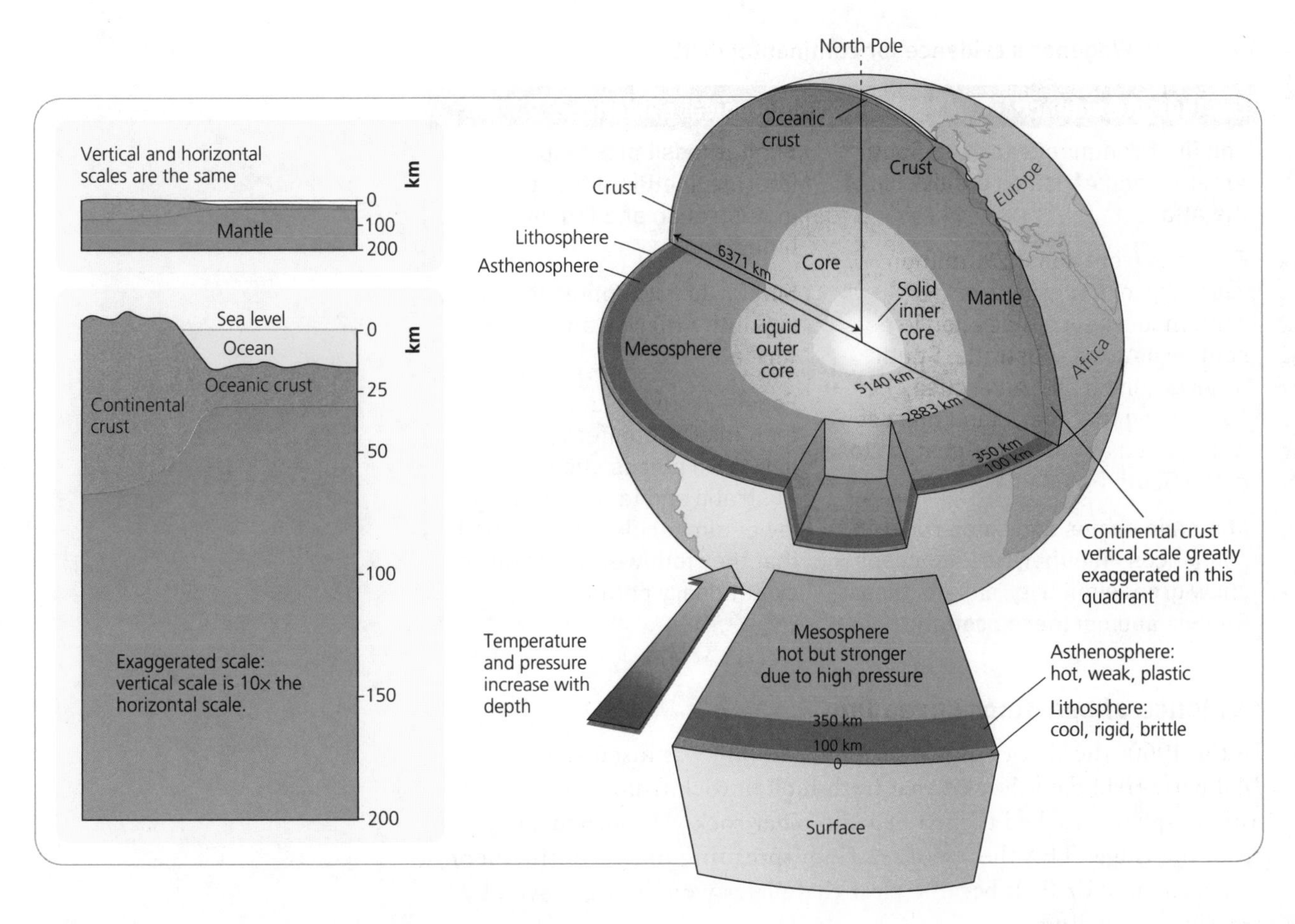

Figure 4.36 The structure of the Earth

- **Mantle:** accounts for more than 80% of the volume of the Earth. It consists of semi-solid rock containing silicon and oxygen, 2,900 km deep. The upper part of the mantle consists of two layers: a layer extending from 100 km to 300 km — the **asthenosphere**. This has plastic properties that allow it to flow under pressure. The layer above this is the **lithosphere**, a rigid layer between the crust and the asthenosphere.
- **Crust:** the outer shell consisting of oceanic crust (solid) composed of dense basalt rock, average 5 km deep, and the continental crust (solid), mainly granite which is less dense than basalt, averaging 30 km deep. Continental crust can be up to 100 km deep under major mountain ranges.

Together the lithosphere and crust make up oceanic and continental plates.

Within the asthenosphere **convection currents**, caused by the intense heat generated in the mantle, mean that this semi-molten layer 'flows', carrying with it the solid lithosphere and crust.

Typical mistake

Tectonic plates are made up of crust and lithosphere, not just crust.

Asthenosphere: the layer in the Earth's mantle below the lithosphere.

Convection currents are the continuous movement of a gas or liquid which when warmed rises and when cooled sinks.

Continental drift is the theory that the continents are mobile.

A **tectonic plate** is a large slab of the Earth's lithosphere and crust.

Sea-floor spreading is the lateral movement of new oceanic crust away from a mid-ocean ridge.

Continental drift and the theory of plate tectonics

In 1912 Alfred Wegener put forward his theory for **continental drift**. He proposed that 250 million years ago, in the Carboniferous period, a large single **tectonic plate** — Pangaea — existed. Initially it broke apart into two land masses to the north and south; this spread continued to give the present-day continental land masses. Evidence for Wegener's theory is summarised in Table 4.28.

Table 4.28 Wegener's evidence for continental drift

Geological evidence	Biological evidence
The fit of continents such as South America and Africa on either side of the Atlantic Evidence from about 290 million years ago of the effects of contemporaneous glaciation in southern Africa, Australia, South America, India and Antarctica, suggesting that these land masses were joined at this time, located close to the South Pole Mountain chains and some rock sequences on either side of oceans show great similarity, e.g. northeast Canada and northern Scotland	Similar fossil brachiopods (marine shellfish) found in Australian and Indian limestones Similar fossil reptiles found in South America and South Africa Fossils from rocks younger than the Carboniferous period, in places such as Australia and India, showing fewer similarities, suggesting that they followed different evolutionary paths

Evidence of sea-floor spreading

In the 1960s the theory of **sea-floor spreading** was discussed. Magnetic field data showed that fresh molten rock from the asthenosphere reached the sea bed and older rock was pushed away from the ridge. This theory of sea-floor spreading linked to the theory of continental drift. It became clear that plates were being moved by sea-floor spreading.

Palaeomagnetism

As lava erupts it cools and the magnetic orientation of iron particles within the lava is 'locked' into the rock. The direction of the Earth's magnetic field changes every 400,000–500,000 years (palaeomagnetism). It was palaeomagnetic data that led to the idea of sea-floor spreading.

The age of sea-floor rocks

In the 1960s an ocean drilling programme showed that the thickest and oldest sediments were near the continents and younger deposits were further out in the oceans, giving further support to the idea of sea-floor spreading.

Evidence from ancient glaciations

In the present day, glacial deposits formed during the Permo-Carboniferous glaciation (about 300 million years ago) are found in Antarctica, Africa, South America, India and Australia. If continents had not moved, then this would suggest that an ice sheet extended from the South Pole to the Equator — this is highly unlikely. If the continents of the Southern Hemisphere are re-assembled near the South Pole, then the Permo-Carboniferous ice sheet shows a much more realistic size.

Fossil records

There are many examples of similar fossils found on separate continents, suggesting the continents were once joined (Table 4.28). If continental drift had not occurred, then either the species evolved independently on different continents, which contradicts Darwin's theory, or they swam to other continents to establish new populations, which is also thought to be unlikely. When the continents of the Southern Hemisphere are re-assembled into the single land mass, the distribution of the fossil types in question forms a more realistic continuous pattern of distribution.

There are distinctive features and processes at plate boundaries

REVISED

Earth's crustal features and processes

The global pattern of plates and plate boundaries

Detailed maps of seismic activity produced in the 1960s showed that most earthquakes were spatially concentrated in narrow bands with large areas generating few earthquakes in between. This led to the idea that the crust and rigid lithosphere were broken into tectonic plates.

- The lithosphere is divided into seven large and three smaller **tectonic plates**.
- The plates are moved by convection currents operating as convection cells within the asthenosphere (Figure 4.37).
- They can move towards each other (**destructive plate margins**), move away from each other (**constructive plate margins**) or slip alongside each other (**conservative plate margins**).
- Most plate movement is slow and continuous but sudden movements produce earthquakes.
- It is at the plate boundaries/margins that most landforms (e.g. fold mountains and volcanoes) are found.

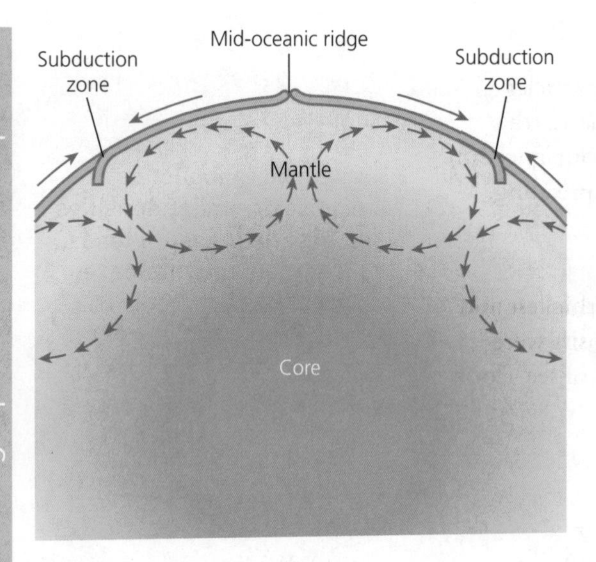

Figure 4.37 How plates move

The features and processes associated with divergent (constructive) plate boundaries

New crust forms at constructive plate margins where rising **plumes of magma** from the upper mantle stretch the crust and lithosphere. The resulting intense volcanic activity builds submarine mountain ranges — **mid-ocean ridges**; parallel faults produce **rift valleys** (Figure 4.38).

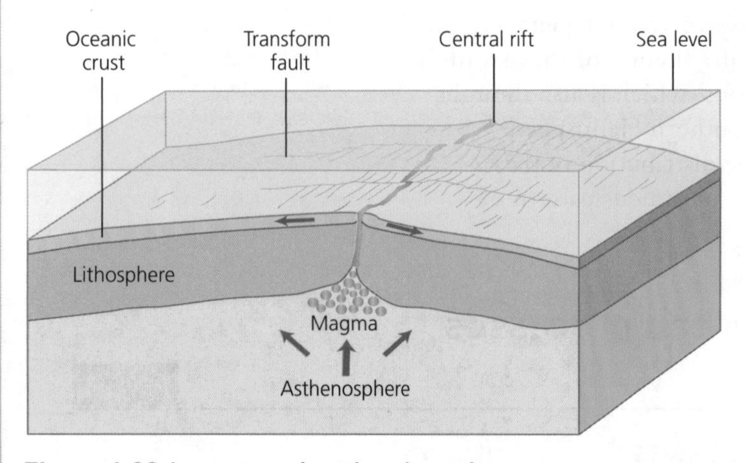

Figure 4.38 A constructive plate boundary

- **Ocean ridges:** formed when plates move apart in oceanic areas. The space between the plates is filled with basaltic lava from below to form a ridge. Volcanoes also exist along this ridge and may rise above sea level, e.g. Surtsey, south of Iceland.
- **Rift valleys:** formed when plates move apart in continental areas. Sometimes the brittle crust fractures as sections of it move and areas of crust drop down between parallel faults to form the valley, e.g. East African Rift Valley.

The features and processes associated with convergent (destructive) plate boundaries

Table 4.29 summarises the features and processes at the three different types of convergent plate boundary.

Exam tip

You should be familiar with the global patterns of plates, their boundaries and their direction of movement.

Exam tip

Learn located examples for the activity at different types of plate boundary.

Table 4.29 Features and processes at convergent plate boundaries

Oceanic–continental plate margins	Oceanic–oceanic plate margins	Continental–continental plate margins
• Different densities of plates • Denser oceanic plate **subducts** under the continental plate • A deep **ocean trench** is formed at the plate boundary • Sediments and rocks fold and are uplifted along the leading edge of the continental plate • Continental crust buckles and mountain chains form, e.g. the Andes • The angle of subduction on the oceanic plate is between 30° and 70°; faulting occurs in the **Benioff zone**, releasing energy in the form of earthquakes	• The slightly denser plate will **subduct** under the other, creating a **trench** • Descending plate melts, magma rises and chains of volcanoes — **island arcs** — form, e.g. the Antilles	• Little if any subduction because of similar densities • Impact and pressure tends to form **fold mountains**, e.g. the Himalayas

The features and processes associated with conservative plate boundaries

At a conservative plate margin, two plates slide past each other. The movement can be violent and an additional build-up of pressure which eventually gives way results in powerful earthquakes. There is no volcanic activity, e.g. San Andreas fault between the Pacific and North American plates.

Revision activity

Draw a series of simple diagrams with annotations to summarise the processes and landforms associated with each type of plate boundary. Figure 4.39 provides an outline example for an oceanic and continental plate (annotations and labels of landforms need to be added).

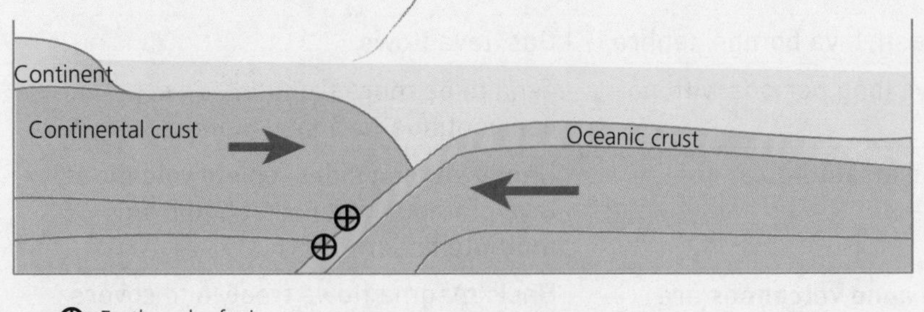

Figure 4.39 Outline diagram for the revision of the processes and landforms at an oceanic–continental destructive plate boundary

Typical mistake

Remember earthquakes and volcanic hazards can occur away from plate boundaries, e.g. earthquakes along fault lines and volcanic activity in Hawaii which is outside a plate boundary.

Exam tip

It is essential to learn key terms for both landforms and processes as this will add clarity and authority to your answers.

Now test yourself

TESTED

62 Describe the structure of the Earth.
63 List the differences between oceanic and continental crust.
64 List **three** pieces of evidence for sea-floor spreading.
65 What is palaeomagnetism?

Answers on p. 224

Exam tip

Practise simple diagrams of the different types of plate boundary and fully annotate them with an explanation of processes and landforms (see Revision activity and Figure 4.39, left).

What are the main hazards generated by volcanic activity?

There is a variety of volcanic activity and resultant landforms and landscapes

REVISED

The range of landforms produced by volcanic eruptions is related to: where and how the eruption takes place; type of lava (**viscosity** affects how the lava flows); materials (ash and lava); and gases produced.

Viscosity describes how well a substance flows.

Different types of volcanoes

Explosive eruptions and effusive eruptions

Volcanic eruptions are divided into two main types: **explosive** — violent because of a build-up of pressure, with viscous magma (e.g. andesite) which prevents escape of gases; and **effusive** — a gentle, free-flowing basic eruption of lava, e.g. basalt. Characteristics are summarised in Table 4.30.

Table 4.30 Characteristics of explosive and effusive eruptions

	Explosive eruptions	Effusive eruptions
Location	Convergent plate boundaries	Divergent plate boundaries
Type of lava	Rhyolite (more acid); andesite (less acid)	Basalt
Lava characteristics	Acid (high % silica), high viscosity, lower temperature at eruption	Basic (low % silica), low viscosity, higher temperature at eruption
Style of eruption	Violent bursting of gas bubbles when magma reaches surface; highly explosive; vent and top of cone often shattered	Gas bubbles expand freely; limited explosive force
Materials erupted	Gas, dust, ash, lava bombs, tephra	Gas, lava flows
Frequency of eruption	Tend to have long periods with no activity	Tend to be more frequent; an eruption can continue for many months
Shape of volcano	Steep-sided strato-volcanoes; caldera	Gently sloping sides, shield volcanoes; lava plateaux when eruption from multiple fissures
Products	**Composite cone volcanoes** are made up of layers of ash and acidic lava. Internal lava flows form **sills** and **dykes**. The acidic magma does not flow easily and solidified magma plugs the vents. **Calderas** (deep craters) form when the cone is destroyed by an explosive eruption	Basic magma flows freely and covers large areas: **flood basalts**

Eruptions not at plate boundaries

These eruptions are associated with **hot spots**. Hot spots are places where a plume of magma rises from the mantle and erupts at the surface. They are usually associated with intense volcanic activity and eruptions of basaltic lava. The Hawaiian chain of volcanic islands is an example.

Size and shape of different types of volcano

Globally there are many volcanoes and a wide variety of eruptions and so they are classified into groups with similar characteristics. Shape and the nature of the eruption are the main criteria on which the classification is based. Figure 4.40 summarises the basic classification.

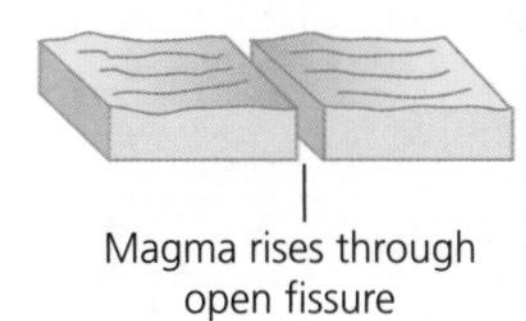

Icelandic lava eruptions are characterised by persistent fissure eruption. Large quantities of basaltic lava build up vast horizontal plains. On a large scale they have formed the Deccan Plateau and the Columbia Plateau.

In **Vulcanian eruptions**, violent gas explosions blast out plugs of sticky or cooled lava. Fragments build up into cones of ash and pumice. Vulcanian eruptions occur when there is very viscous lava which solidifies rapidly after an explosion. Often the eruption clears a blocked vent and spews large quantities of volcanic ash into the atmosphere.

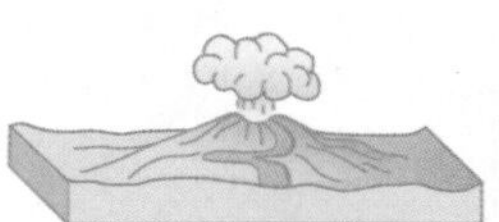

Hawaiian eruptions involve more noticeable central activity than the Icelandic type. Runny, basaltic lava travels down the sides of the volcano in lava flows. Gases escape easily. Occasional pyroclastic activity occurs but this is less important than the lava eruption.

Vesuvian eruptions are characterised by very powerful blasts of gas pushing ash clouds high into the sky. They are more violent than Vulcanian eruptions. Lava flows also occur. Ash falls to cover surrounding areas.

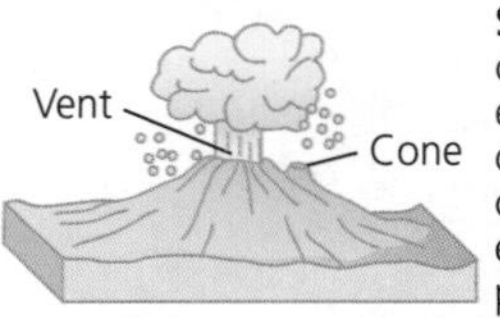

Strombolian eruptions are characterised by frequent gas explosions which blast fragments of runny lava into the air to form cones. They are very explosive eruptions with large quantities of pyroclastic rock thrown out. Eruptions are commonly marked by a white cloud of steam emitted from the crater.

In a **Plinian eruption**, gas rushes up through sticky lava and blasts ash and fragments into the sky in a huge explosion. The violent eruptions create immense clouds of gas and volcanic debris several kilometres thick. Gas clouds and lava can also rush down the slopes. Part of the volcano may be blasted away during the eruption.

Figure 4.40 Shape classification of different volcanoes

Super-volcanoes are volcanoes that erupt more than 1,000 km^3 of material in a single event. An example is the Yellowstone super-volcano in Wyoming, which has a caldera measuring 75 km in diameter.

The volcanic explosive index

Magnitude (the amount of material erupted) and **intensity** (the speed at which the material is erupted) can be used to compare different eruptions. The **Volcanic Explosivity Index** (**VEI**) combines these two factors into a single figure on a scale of 0 (least explosive) to 8 (most explosive). Each increase in number represents a ten-fold increase in explosivity.

Exam tip

The key to understanding eruption behaviour is the viscosity of the magma and the extent of the release of gases and steam.

Volcanic eruptions generate distinctive hazards

REVISED

Different types of volcanic eruptions and different types of hazards

Hazards generated by volcanic eruptions

There are different types of volcanic hazards (see Figure 4.41).

- **Lava flows:** flows or streams of molten rock that pour from an erupting vent. The speed at which lava moves depends on the type of lava (**basaltic lava** is free-flowing and runs considerable distances, while **acidic lava** is thick and does not flow easily), its viscosity, the steepness of the ground and whether the lava flows as a broad sheet, through a confined channel, or down a lava tube. Lava of any type is extremely destructive and will burn, bury or bulldoze infrastructure, property, natural vegetation and agricultural land.

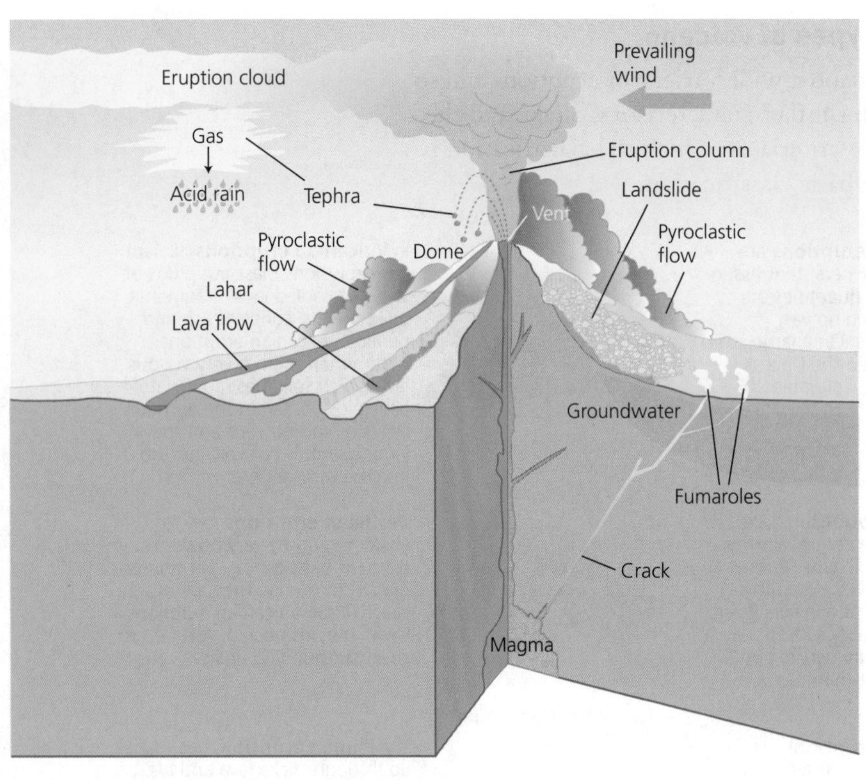

Figure 4.41 Types of volcanic hazard

- **Pyroclastic flows — nuées and ardente:** flows of gas and tephra which are extremely hot (over 500°C) and flow down the side of the volcano at speeds of over 100 km/h.
- **Gas emissions:** carbon dioxide, carbon monoxide, sulphur dioxide and chlorine escape through fumaroles (openings in or near a volcano, through which hot sulphurous gases escape). When sulphur dioxide combines with water, acid rain is produced, leading to weathering and pollution.
- **Tephra:** volcanic bombs and ash ejected into the atmosphere. Size ranges from ash to larger bombs of > 6 cm diameter.
- **Lahars and flooding associated with melting ice:** lahars are a type of mud flow. Snow and ice on a volcano summit melt during the eruption. Rock, ash and soil mix together and destroy and bury anything in the path of the rapid flow of material as they follow valleys. The melting of snow and ice associated with volcanic eruptions can also lead to flooding as large volumes of water are released. In Iceland, these are known as **jökulhlaups**.
- **Tsunamis:** violent eruption of an island volcano can displace oceanic water and lead to a tsunami — a large wave travelling at speeds up to 600 km/h.

Typical mistake

Molten rock beneath the Earth's surface is magma; at the surface this is known as lava.

Revision activity

Volcanoes can also be classified according to the type of eruption. Construct a table to summarise the main types of eruption, e.g. Icelandic lava eruptions and Vesuvian eruptions.

Now test yourself

TESTED

66 Outline the differences between explosive and effusive volcanic eruptions.
67 Describe **four** types of volcanic hazard.

Answers on p. 224

What are the main hazards generated by seismic activity?

There is a variety of earthquake activity and resultant landforms and landscapes

REVISED

An earthquake is a release of stress in the Earth's crust. A series of shock waves originate from the earthquake **focus** (the location where the stress is released) and on the Earth's surface this point is known as the **epicentre**. Fore-shocks can be released before the main earthquake event. The main locations for earthquake activity are mid-ocean ridges, ocean trenches and island arcs, collision zones and conservative plate margins.

Earthquake characteristics

Exam tip

It is useful to have a basic knowledge of seismic waves. There are three types: primary (fast, low frequency), secondary (half the speed of primary, high frequency) and surface (slow, low frequency).

Shallow-focus earthquakes

- Surface down to approximately 70 km.
- Often occur in brittle rocks.
- Generally release low levels of energy but high-energy shallow quakes can cause severe impacts.

Deep-focus earthquakes

- 70–700 km.
- Increasing depth leads to high pressure and temperature.
- Less frequent but very powerful.
- Full understanding of deep-focus earthquakes is evolving; water and change in minerals may be contributing factors.

Measuring earthquake magnitude

- **Richter scale:** developed in 1935; uses the amplitude of seismic waves to measure magnitude. Scale is logarithmic from 1 to 9 (although there is no upper limit); each whole number increase is a ten-fold increase.
- **Moment Magnitude scale:** scale of 1.0–9.0, measuring energy release as related to geology, the area of the fault surface and the amount of movement on the fault. Accurate for large earthquakes as it uses the physical movement caused by the earthquake. It is not used for small earthquakes.
- **Modified Mercalli scale:** measures earthquake intensity and impact. It relates to impacts felt and seen by those affected — it is qualitative not quantitative.

Exam tip

Magnitude is only one of several factors that will affect the impact of a hazard. In some instances the relationship may be quite weak.

The effects of earthquakes on landforms and landscapes

Earthquakes have had a dramatic and widespread effect on global landscapes across different geological timescales. Mountain ranges have been created (e.g. the Himalaya–Karakoram Range in Asia), as well as major fault systems, **rift valleys** (e.g. East Africa) and **escarpments**.

A **rift valley** is a valley formed by downfaulting between parallel faults.

An **escarpment** is a tilt block forming an extensive upland area, with a short, steep, scarp slope and a long, gentle, dip slope on the other side.

Earthquakes generate distinctive hazards

REVISED

Hazards generated by earthquakes

Earthquakes become hazards when they interact with people. The main hazards generated by earthquakes are summarised in Table 4.31.

Table 4.31 Earthquake hazards

Ground shaking and ground displacement	This is the vertical and horizontal moving of the ground. Severity depends on the earthquake magnitude, distance from the epicentre and geology
Liquefaction	When violently shaken, soils with a high water content lose their mechanical strength and become fluid
Landslides and avalanches	Slope failure as a result of the ground shaking
Tsunamis	A tsunami is a giant sea wave generated by shallow-focus underwater earthquakes (also volcanic eruptions and large landslides into the sea). Tsunamis have long wavelength (often over 100 km) and low wave height (under 1 m) in the open ocean. They travel quickly (speeds over 700 km/h) but on reaching the shallow water bordering land they increase in height. A wave trough forms in front of the tsunami where sea level is reduced: this is called a **drawdown**. Behind this comes the tsunami itself, sometimes as high as 25 m or more
Flooding	Earthquakes can indirectly cause flooding in a number of ways: triggering tsunamis, destabilising/destroying dams, destroying and/or lowering protective levees

Exam tip

Tsunamis present an exceptional hazard as effective protection is extremely difficult and the impact is often a long way from its origin.

What are the implications of living in tectonically active locations?

Despite the knowledge that certain areas (mostly associated with plate boundaries) have the potential for earthquakes and volcanic eruptions, millions of people continue to live in such locations for a range of sociocultural, economic and environmental reasons, e.g.:

- Weathered lava produces fertile soils in countries such as Japan and Indonesia.
- Volcanoes provide opportunities for economic activity such as tourism, e.g. in Iceland and Italy.
- Tectonically active areas produce geothermal power, e.g. Iceland and Indonesia.
- Volcanic eruptions supply minerals such as sulphur, used in a variety of industries.

Now test yourself

TESTED

68 What is liquefaction?
69 What is drawdown?

Answers on p. 224

There are a range of impacts people experience as a result of volcanic eruptions

REVISED

One of the contributing factors to people living in locations where there is a volcano is the description of volcanoes as active, dormant or extinct. Clearly there is a different risk perception for areas with an active rather than an extinct volcano. However, the classification is rather simplistic and open to question regarding time perspectives and the definition of active.

Exam tip

When writing about a hazard event the impact is the result of exposure and vulnerability of people, buildings and infrastructure.

Case study of two countries at contrasting levels of economic development to illustrate the impacts people experience as a result of volcanic eruptions

Revision activity

Make summary notes, based on the structure of the case study of earthquake activity in Nepal provided online (see p. 3), of **two** case studies covered in class to revise how countries at contrasting levels of economic development are impacted by volcanic eruptions. Remember to focus on:

- the reasons why people choose to live in a tectonically active location
- the impacts people experience as a result of volcanic eruptions
- economic, environmental and political impacts on the country

There are a range of impacts people experience as a result of earthquake activity

REVISED

Case study of two countries at contrasting levels of economic development to illustrate the impacts people experience as a result of earthquake activity

Revision activity

Online (see p. 3) you will find a case study on Nepal, illustrating a country living with earthquakes. Make summary notes of a case study you have covered in class, to revise how a country at a contrasting level of economic development to Nepal has been impacted by earthquake activity. Use the online case study to help you with structure. Remember to focus on:

- the reasons why people choose to live in a tectonically active location
- the impacts people experience as a result of earthquake activity
- economic, environmental and political impacts on the country

Now test yourself

TESTED

70 What are the differences between shallow- and deep-focus earthquakes?

71 Describe **four** types of hazard generated by earthquakes.

Answers on p. 224

What measures are available to help people cope with living in tectonically active locations?

As a result of people living in locations affected by earthquakes and volcanoes, there is a need to develop strategies to manage and cope with such hazards. These include mitigation and adaptation strategies, together with assessment of risk and resilience. Table 4.32 summarises some strategies for managing tectonic hazards.

Table 4.32 Some strategies for managing tectonic hazards

Modify the event	Modify people's vulnerability	Modify people's loss
Not possible for the vast majority of volcanic eruptions. However, the following have been tried with some success: • lava-diversion channels • spraying lava to cool it so it solidifies • slowing lava flows by dropping concrete blocks Earthquakes: • nothing can be done to modify an earthquake event	**Education:** recognise signs of possible eruption; what to do when an eruption occurs, e.g. evacuation routes; drills to practise what to do when a tectonic event strikes, e.g. in an earthquake, get to open space away from buildings or shelter under a table in a doorway **Community preparedness:** e.g. building of tsunami shelters and walls; strengthening of public buildings, e.g. hospitals, fire stations, schools **Prediction and warning:** increasing use of technology to monitor particularly active locations, e.g. individual volcanoes **Hazard-resistant building design:** e.g. cross-bracing of buildings to support them during an earthquake; steep-sloping roofs to prevent ash building up **Hazard mapping:** e.g. predicted lahar routes; ground likely to liquefy in an earthquake **Land-use zoning** to avoid building in locations identified by hazard mapping	Emergency aid, e.g. bottled water, medical supplies, tents, food packs Disaster-response teams and equipment, e.g. helicopters and heavy lifting machinery Search and rescue strategies Insurance for buildings and businesses Resources for rebuilding public services, e.g. schools and hospitals, and help for individuals to rebuild homes and businesses

Aseismic design features include:

- stepped building profile
- varied building height
- angled windows
- soft storey
- reinforcements, e.g. bracing, steel frames and deep foundations
- consideration of building site to avoid difficulties such as fault lines, soft soil and slopes

Revision activity

Produce a table that summarises some of the general strategies to manage hazards from earthquakes and volcanic activities.

Exam tip

ACs are more able to invest in the most advanced protection for seismic events. Often expertise is sent from ACs to LIDCs.

There are various strategies to manage hazards from volcanic activity

REVISED

Case study of **two** countries at contrasting levels of economic development to illustrate strategies used to cope with volcanic activity

Online (see p. 3) you will find a case study of Italy to illustrate strategies used to cope with volcanic activity.

Revision activity

Make summary notes of a country at a contrasting level of economic development to Italy to show strategies used to cope with volcanic activity. Use the online case study to help you with structure. Remember to focus on:

- attempts to mitigate against the event
- attempts to mitigate against vulnerability
- attempts to mitigate against losses

There are various strategies to manage hazards from earthquakes

REVISED

Case study of **two** countries at contrasting levels of economic development to illustrate strategies used to cope with hazards from earthquakes

Revision activity

Make summary notes, based on the structure of the case study of volcanic activity in Italy provided online (see p. 3), of two countries at contrasting levels of economic development to show strategies used to cope with hazards from earthquakes. Remember to focus on:

- attempts to mitigate against the event
- attempts to mitigate against vulnerability
- attempts to mitigate against losses

Now test yourself

TESTED

72 Outline **two** possible responses to mitigate the impact of earthquakes.

Answers on p. 224

The exposure of people to risks and their ability to cope with tectonic hazards changes over time

REVISED

How and why have the risks from tectonic hazards changed over time?

Change in the frequency and impact of tectonic hazards over time

The number of natural disasters has increased over time but the increase in earthquakes and volcanic eruptions is less pronounced. There has been an increase over the past 50 years; on average there are now 30 volcanic eruptions and earthquakes a year. Data show that earthquakes have a significantly greater impact in terms of number of deaths than volcanic eruptions.

Generally, the greater the magnitude of an eruption or earthquake, the less frequently it occurs. Recurrence intervals indicate that high-magnitude events recur over longer periods of time. High-magnitude, rare events can release large amounts of energy and have the greatest impact on human populations.

The disaster risk equation

The disaster risk equation gives an idea of the hazard vulnerability of a location. It is expressed as:

$$\text{Risk } (R) = \frac{\text{Frequency of magnitude of hazard } (H) \times \text{Level of vulnerability } (V)}{\text{Capacity of the population to cope and adapt } (C)}$$

$$R = \frac{H \times V}{C}$$

The relationship between the magnitude and impact of a tectonic event can be influenced by how often an event occurs and the time interval between events. The most vulnerable are those who experience a wide-ranging impact from a small-scale event.

Possible future strategies

- Effectiveness of current strategies needs to be assessed and evaluated in order to draw lessons for the future.
- It is not currently possible to make predictions of when and where earthquakes will happen. A specific and reliable 'precursor' (a characteristic pattern of seismic activity or a physical, chemical or biological change) is needed in order to confidently predict an earthquake event within a specific time and place. So far, the search for such a precursor has been unsuccessful. Most geoscientists do not believe that there is a realistic prospect of accurate prediction in the foreseeable future.
- Future research will therefore focus on improving the forecasting of earthquakes.
- The UN identifies the alleviation of poverty as a priority in reducing the effects of earthquakes in the future.
- Falling buildings are by far the greatest cause of casualties during earthquakes, so it is essential that building design to withstand tectonic events continues to make advances in the future.

The relationship between disaster and response

Characteristic human responses to hazards

- Responses occur at different levels: individual, community, national government and international.
- Resilience is the sustained efforts of communities to respond to and withstand the effects of hazards.
- Hazards are also managed by the integration of prediction, prevention and protection plans.
- Prediction is not always possible scientifically but, with careful monitoring, warnings can be issued.
- Natural hazards cannot be prevented but some of the dangerous secondary impacts can be controlled, e.g. lava flows can be diverted.
- Protection aims to minimise the impact of a hazard event. This usually involves adaptations to the built environment, e.g. earthquake-proof buildings, sea defences.
- Risk sharing involves public education and awareness of the measures available to reduce impact, e.g. emergency responses and evacuation procedures.
- All of the above can be put together in a **hazard management cycle**, as shown in Figure 4.42.

Figure 4.42 The hazard management cycle

> **Exam tip**
>
> The Park model will be affected by the magnitude of the event, the speed of onset, human monitoring, preparedness and the quality and quantity of relief.

The Park model of human response to hazards

The Park model shows that hazards have varying impacts over time: before the disaster; when the event happens; and post-event relief (rescue), rehabilitation (restoring the functioning of public services) and reconstruction (rebuilding the public and economic system, replacing infrastructure and governance). The disaster-response curve (Park, 1991), shown in Figure 4.43, shows how the effect of a disaster on people can be generalised.

Now test yourself TESTED

73 State **two** ways in which risks from hazards have changed over time.
74 State **two** possible future strategies to deal with seismic events.

Answers on p. 224

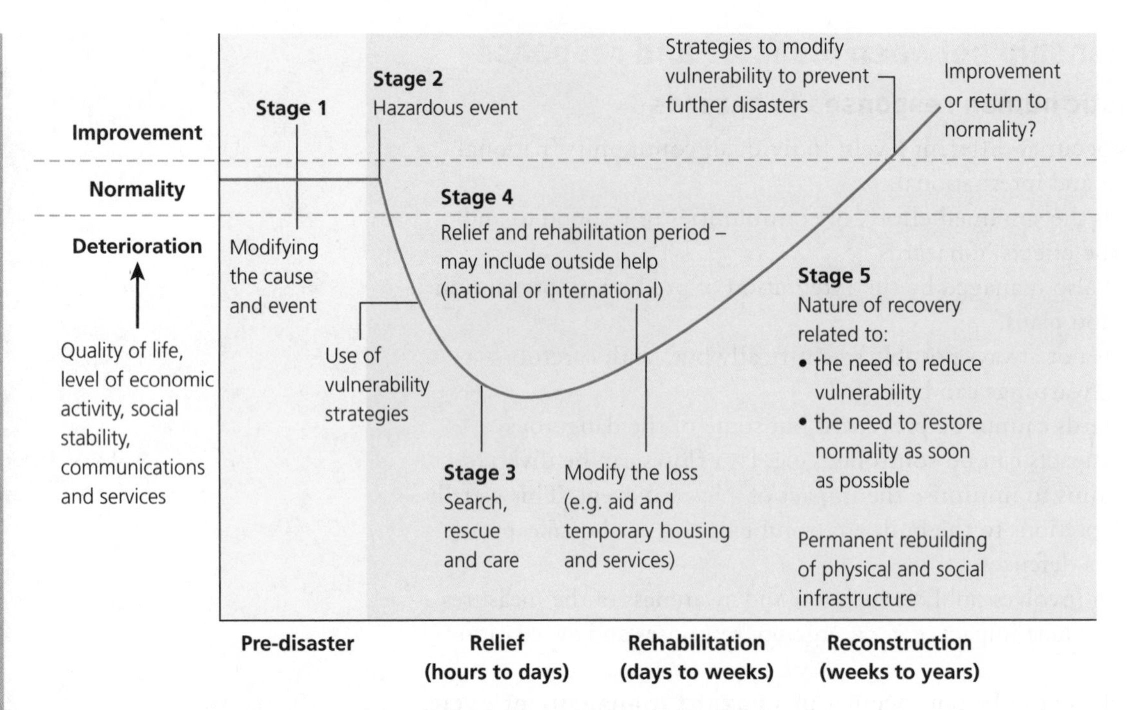

Figure 4.43 A disaster-response curve (after Park, 1991)

Now test yourself

TESTED

75 When does a natural hazard become a natural disaster?
76 What factors determine the impact of a hazard?
77 What factors influence hazard perception?
78 What is meant by hazard management?

Answers on p. 224

Summary

- Various forms of evidence exist for the theory of continental drift and plate tectonics, including geological and biological evidence.
- A range of distinctive features and processes exist at constructive, destructive and conservative plate boundaries.
- Volcanic and seismic activity generate a range of different landforms, landscapes and hazards.
- Despite the risk, people continue to live in tectonically active areas and this has impacts on people, politics, the economy and the environment.
- A number of strategies and measures are available to help cope with the risks generated from tectonic activity.
- These strategies and measures vary across a range of countries at contrasting levels of economic development.
- The exposure of people to risks and their ability to cope with tectonic hazards change over time, particularly with regard to the frequency and impact of the hazard event and the degree of exposure.
- Possible future strategies to cope with tectonic hazards exist.
- The Park model sets out a theory of the relationship between disaster and response.

Exam practice

21 Explain how higher viscosity magma can cause explosive volcanic eruptions. [4]

Table 4.33 Some of the most powerful and deadly earthquakes since 1900 (Wiki commons)

Country	Year	Magnitude
Chile	1960	9.4–9.6
Alaska	1964	9.2
Indian Ocean Sumatra	2004	9.1–9.3
Pacific Ocean Japan	2011	9.1
Russia	1952	9.0
Chile	1922	8.5
China	1920	8.6
Turkmenistan	1948	7.3
Turkey	1999	7.2
India	2001	7.7

22 Identify **three** limitations of the data evidence in Table 4.33. [3]

23 Explain the evidence for continental drift and plate tectonics. [6]

24 Examine how tectonic activity can influence place identity. [8]

25 'Earthquakes present a greater disaster risk than volcanoes.' How far do you agree with this statement? [12]

Answers and quick quiz 4E online

ONLINE

Now test yourself answers

Chapter 1 Landscape systems

Option A Coastal landscapes

1 Because they allow the transfer of both energy and matter. Closed systems transfer just energy with outside space.

2 When balance is maintained by adjustments to inputs and outputs; the system is constantly changing.

3 Energy from waves, tides, wind and sea currents. Sediment, geology, thermal energy from the sun and sea-level change.

4 A process by which waves break onto an irregular-shaped coastline. Approaching a shoreline the waves drag in shallow water, the wave becomes steep and short, and the part of the wave in the deeper water moves faster. The wave bends, the low-energy wave spills into the bay and most of the energy is concentrated on the headland.

5 Ocean currents are generated by the Earth's rotation and are set in motion by the wind. Warm ocean currents transfer heat from low to high latitudes and cold ocean currents from high to low latitudes.

6 A sediment budget is the balance between the input and output of sediment on a coastline. Calculation is complex as all possible inputs and outputs need to be identified.

7 Because they loosen material and make it available for use in processes such as abrasion; also weathering exposes surfaces for further erosion.

8 Geology: hard and soft rock; climate: affects weathering processes; human activity.

9 The coastline will have characteristic features of cliffs, wave-cut platforms, caves, arches and stacks. There will be rugged cliff faces and no long, extended beaches.

10 Weathering processes (mechanical, biological and chemical) weaken rock in situ. Erosion processes attack the cliff face and remove and use the weathered material — hydraulic action, wave quarrying, abrasion, attrition, solution.

11 Flatter beach profiles are the result of high-energy, destructive waves and steep beach profiles of low-energy, constructive waves.

12 From pioneer species such as cord grass, which slows water movement and stabilises mud, to species such as marsh grass, which develop soil, to climax vegetation of ash, alder and oak that establish in less saline conditions.

13 Cusps are small, semi-circular depressions; they are temporary features formed when a collection of waves reach the same point and swash and backwash are of equal strength.

14 High-energy coastline features include: strong wave power, high waves, strong prevailing wind, long length of fetch, erosion greater than deposition.

15 Beaches, spits, tombolos, on-shore and off-shore bars.

16 Formed when beaches and cliff lines are elevated above sea level and the cliffs are exposed following deglaciation and the unloading of ice sheets. They are emergent landforms.

Option B Glaciated landscapes

17 The variation in the glacial mass balance. The variation between inputs in the zone of accumulation and outputs in the zone of ablation.

18 In winter accumulation will exceed ablation leading to a positive budget, and in summer ablation will exceed accumulation leading to a negative budget.

19 Erosion, transportation and deposition processes by wind.

20 It affects the impact of weathering, erosion and mass movement processes. Structure, permeability and angle also affect weathering and geomorphic processes.

21 An aspect facing the sun will lead to more melting and a negative budget, while an aspect facing away from the sun will mean less melting and a positive budget.

22 A layer of compact snowflakes.

23 Warm-based glaciers occur in temperate areas. They are small, with summer melts. Lubrication means more movement and more erosion, transportation and deposition. Cold-based glaciers occur in polar areas. They comprise large glaciers and vast ice sheets. All ice in cold-based glaciers is below melting point, meaning slower movement, and the glacier is often frozen to the bed, resulting in less erosion, transportation and deposition.

24 Warm-based glaciers: rapid movement — slippage, creep and bed deformation. Cold-based glaciers: slow movement, intergranular flow and laminar flow.
25 Processes operating underneath snow patches: freeze–thaw, chemical weathering and transport by meltwater lead to the formation of nivation hollows.
26 Supra-glacial, en-glacial and sub-glacial transportation.
27 Till is angular, unsorted and unstratified rock deposited by the ice. Outwash is material deposited by meltwater; it is coarse near the glacier and finer when carried across the outwash plain.
28 Abrasion: material in the glacier rubbing away at rock surfaces. Plucking: the glacier freezes onto and into the rock surfaces. As it moves, it takes away rock fragments that have been previously loosened by freeze–thaw action.
29 (a) Drumlin: a low, streamlined hill deposited and shaped by a moving ice sheet.
(b) Lateral moraine: derived from frost shattering of the valley sides; it is carried at the edge of the glacier.
30 A very long, narrow ridge of sorted, stratified, coarse sand and gravel.
31 Glacio-fluvial deposits will be rounder and more sorted; ice deposits are angular and unsorted.
32 Freeze–thaw weathering.
33 Solifluction: summer meltwater in the upper layers of permafrost cannot drain away, and the lubrication means that soil is moved on the most gentle of slopes.

Option C Dryland landscapes

34 Solar radiation, wind and atmospheric humidity.
35 A numerical index of dryness. Mean annual precipitation is divided by mean annual potential evapotranspiration.
36 An area of below-average precipitation situated in the lee of an upland.
37 It has a localised influence; temperature falls with altitude; there are large temperature ranges, low humidity and low precipitation.
38 Southerly aspects lead to warmer temperatures and higher rates of evapotranspiration leading to more dryness.
39 A treeless region in sub-Arctic and high mountain environments which has a short growing season and severe winter temperatures.
40 Summer meltwater in the upper layers of permafrost cannot drain away; the lubrication means that soil is moved on the gentlest of slopes.
41 Polar drylands, mid- and low-latitude deserts and semi-arid environments.
42 Short-lived and flows only after heavy rain.
43 Creep, saltation and suspension.
44 Wind erosion is the abrasive effect of sand grains close to the land surface. Wind transport is the saltation and creep of sand grains. Deflation is the removal of material by wind.
45 Linear (seif), star, parabolic, crescent.
46 Zeugen is a collective term for rock pillars, rock pedestals and yardangs which have been undercut where less resistant rock underlies a layer of resistant rock.
47 Frost heave is the downslope displacement of soil particles that results from cycles of freeze–thaw.
48 Jointed, permeable (both allow water in), soft (easily shatters).

Chapter 2 Earth's life support systems

1 For processes such as photosynthesis, respiration and transpiration, to transport minerals from the soil and to maintain structure.
2 The atmosphere, oceans, rocks, soil and the biosphere.
3 Evaporation, transpiration, condensation, cloud formation, precipitation, cryospheric processes.
4 Photosynthesis, respiration, oxidation and weathering.
5 It is the long-term balance between inputs and outputs in a drainage basin system. It is expressed in the following equation, where P is precipitation, Q is run-off, E is evapotranspiration and S is storage:

$$P = E + Q \pm S$$

6 In an open system there is the transfer of energy and matter, in a closed system just energy.
7 (i) Water evaporates from the surface of the Earth and condenses around nuclei to form visible water droplets: clouds.
(ii) Water vapour evaporates into the atmosphere; condensation occurs when air temperature reaches its dew point or due to adiabatic cooling: leading to rainfall.
8 Temperature: warmer temperatures lead to higher rates of evapotranspiration as warm air can hold more water vapour. Wind: evapotranspiration increases as wind moves humid air away and the air does not become saturated as quickly. Humidity: the more humid it is the lower the evapotranspiration as the air becomes saturated quickly.

9 On sunny days, the air is heated by warm surfaces; it rises rapidly, cools, condenses and forms convectional rainfall. The rainfall is in short, heavy bursts because of the rapid process of heating and uplift.

10 Combustion, respiration and diffusion.

11 Organic matter which can be vegetation or fossil fuel.

12 Carbon dioxide moves from the atmosphere to the oceans by diffusion. At low latitudes, warm water absorbs carbon dioxide; at high latitudes where cold water sinks, the carbon is transferred deep into the ocean. Where the cold water returns to the surface and warms again it loses carbon dioxide to the atmosphere; in this way carbon dioxide is in constant exchange between the oceans and the atmosphere — the oceanic carbon pump.

13

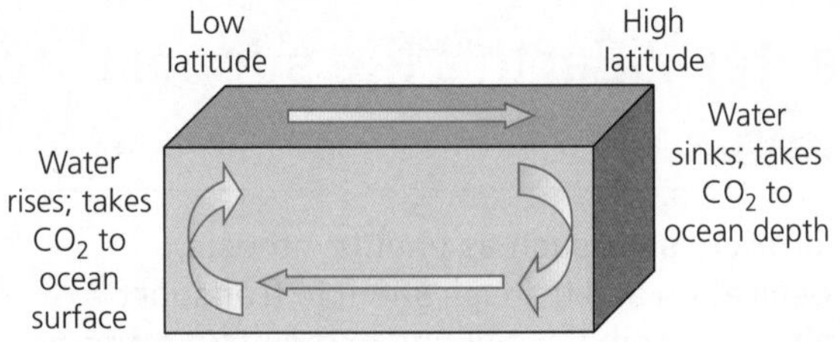

Low latitude

High latitude

Water rises; takes CO_2 to ocean surface

Water sinks; takes CO_2 to ocean depth

14 Decomposition is faster in warm climates as there is more bacterial activity than in cold climates.

15 A natural or artificial process by which carbon dioxide is removed from the atmosphere and stored.

16 It has an important positive feedback loop: a warm atmosphere = more evaporation and more water vapour in the atmosphere = more absorption of thermal energy radiated from the Earth = a warmer atmosphere, and so on. Water vapour controls the Earth's temperature and accounts for a high proportion of the warming effect (some scientists state 60%).

17 Acidification limits the capacity of the oceans to store carbon.

Chapter 3 Changing spaces; making places

1 Location, locale and sense of place.

2 Space exists between different places and does not have the same meanings.

3 Age, gender, sexuality, religion, role.

4 Memory and emotional attachment.

5 Through the process of globalisation, time–space compression and global TNCs making high streets look the same.

6 By the glocalisation process, the promotion of local goods and services and the adaptation of global products to reflect local identities.

7 The lagging behind of members of society in a number of related aspects of life.

8 The most frequently used measure is the Multiple Deprivation Index which is a composite index of income, employment, health, education, crime, living environment and access to housing and services.

9 Wealth, housing, health, education, access to services.

10 Impact of TNCs on people's lives: through employment, living standards, quality of life, wellbeing, skills training, quality of the environment.

11 The absolute or relative decline in the importance of manufacturing in the economy of a country or region.

12 Because many had an employment base in manufacturing industries, e.g. Birmingham: car manufacture; Glasgow: shipbuilding.

13 New or expanding economic activity creates employment; as there is more affluence from employment there is a growth in goods and services; more money is spent in the local economy; as the wealth of an area increases it stimulates more economic activity and inward investment, and so on.

14 (i) Rebranding is development aimed at changing negative perceptions of a place and making it more attractive to invest in.

(ii) Reimaging is development associated with rebranding, usually involving cultural, artistic and sporting elements.

(iii) Regeneration is the investment of capital and ideas into an area to revitalise and renew its socio-economic and environmental status.

15 Top-down rebranding is led by large-scale organisations, e.g. planning departments of local authorities and development agencies.

Option A Trade in the contemporary world

16 Europe, North America and Asia show the highest levels of trade. The Middle East is also well connected to trade flows. Although South America and Africa are connected with trade movements, Africa has the lowest levels, for example between North America and Africa 0.8% and Africa and Asia 1.7%.

17 High levels of merchandise trade between Europe and Asia and low values with Africa. Asia has nearly ten times the trade of Africa. Europe also dominates. Reasons include investment by

governments in Europe, Asia and North America in education, training and skills development. Attraction of cheap labour in Asia leads to foreign investment. Europe, Asia and North America are the largest global markets for merchandise because of high levels of economic development and consumer demand.

18 FDI is used to fund development and development projects. MNCs locate their manufacturing operations in countries where there is cheap labour and/or an abundance of resources; this is often in developing countries.

19 The concept that countries or regions benefit from specialising in economic activity in which they are relatively more efficient and skilled.

20 Trade increases employment. Through the upward multiplier effect this leads to economic growth as there is more affluence, more spending and more investment by the government. Trade also leads to infrastructure development, which leads to more economic development. As countries develop and their investment in business, education and training increases, further investment takes place from within the country and beyond, furthering overall economic development.

21 (a) Conflict can occur as all countries in a trading situation want to secure the best deal for their citizens and businesses. Conflict can also arise when countries are accused of promoting 'protectionist' policies, e.g. the USA holding back on some regional trade agreements.

(b) Displacement of communities following land-grabbing investments, use of child labour, unfair trade rules such as tariffs which can adversely affect small-scale operations.

22 Transport is now more efficient because of the increased size and standardisation of containers. Freer movement of goods resulting from reduction in costs and computerised logistics has led to more efficient handling and more trade.

23 Outsourcing is a cost-saving strategy where a company that has comparative advantage provides goods or services for another company even though they could be produced in-house.

24 The global reorganisation of production because of deindustrialisation in ACs and the spread of MNCs.

25 Higher-paid jobs tend to be in ACs and lower-paid manufacturing jobs in LIDCs.

Option B Global migration

26 The difference between people moving permanently into an area and out of that area.

27 Funds transferred from migrants back to their home country.

28 Because they increase spending power and affluence in the country receiving the remittances and through the upward multiplier this creates economic growth. Remittances are also a source of foreign exchange.

29 New skills learnt by returning migrants while they were working away can contribute to economic growth in their home country. GDP and tax base of host nations are boosted by working migrants. Migrants can stimulate demand and growth in the host nation. Migrants can fill skills gaps allowing a host nation to develop.

30 A growing independence, status and freedom of female migrants. Increasing numbers of highly skilled women workers from Africa and Latin America. Less discrimination in the labour markets of Europe and the USA increases female migrant flows into these regions.

31 More interconnected global economies in the South. A number of fast-growing economies in the South. Increasing cost of South to North migration.

32 Increasing numbers of international conflicts lead to a growing desperation to move.

Option C Human rights

33 Humanitarian (by a state or group of states). Intervention by a global organisation such as the UN, economic sanctions, prosecution of individuals responsible for human rights violations.

34 The way in which geographical factors shape international politics.

35 Gender inequality, sexual exploitation, bonded labour.

36 Social media can help the flow of ideas. Remote sensing and satellite imagery are used in surveillance in dangerous areas.

37 Norms are moral principles, customs and ways of living that are universally accepted as standard behaviour.

38 In a positive way through medical assistance, provision of food and shelter and military protection. In a negative way by damage to property and infrastructure, population displacement, civilian casualties, disruption to education and dependence on aid, e.g. food aid can undermine the local agricultural economy.

39 Prevention of casualties and provision of safety and protection for aid workers.

Option D Power and borders

40 A nation is a large group of people with strong bonds of identity. A state is an area of land of an independent country with well-defined boundaries. A nation may be one country or its people may live across adjoining countries or even spread globally. States often contain several national groups.

41 Internal sovereignty is a state's exclusive authority. External sovereignty is mutual recognition among sovereign states.

42 Sovereignty is the absolute authority which independent states exercise in the government of the land and people in their territories. Territorial integrity is when states exercise their sovereignty within a specific territory.

43 Norms are moral principles, customs and ways of living that are universally accepted as standard behaviour.

44 The way in which global affairs are managed.

45 It undermines the sovereignty of nations and it marginalises poorer countries.

46 Business decisions that impact many people are taken outside the host country. TNCs may challenge human rights within the host country.

47 They challenge a state when they have strong identities and may demand independence. They may also create internal conflicts between ethnic groups which destabilise the state.

48 The Earth's shared natural resources, e.g. the oceans and the atmosphere.

49 Global organisations such as the UN require vast monetary resources to fund their projects and pay thousands of staff.

50 (a) Agricultural training to improve food security; building of democratic institutions; education; technical assistance to improve legislation; support of a fair electoral process; upholding human rights; integrating gender equality into systems.

(b) Security and protection of civilians; negotiate a period of ceasefire; monitor human rights; early warning of conflict; reduce forced conscription of child soldiers; strengthen the rule of law; assistance for IDPs and their return.

Chapter 4 Geographical debates options

Option A Climate change

1 Glacials are cold climatic periods (ice ages) lasting tens of thousands of years. Interglacials are shorter spells of warmer climate when ice sheets and glaciers retreat.

2 The current geological period when humankind is the main driver of environmental change.

3 External forcings: astronomical shifts in the Earth's orbit and axial tilt, equinoxes and fluctuations in the solar output. Internal forcings: volcanic eruptions, continental drift, changes in ocean circulation and fluctuations in the atmospheric carbon dioxide.

4 Increases in global temperatures, shrinking valley glaciers and ice sheets, rising sea level, increased atmospheric water vapour, decreasing snow cover and sea ice.

5 Satellite imagery and ground photographs.

6 The reflecting of sunlight from the Earth's surface.

7 The greenhouse effect is the natural warming influence of gases such as carbon dioxide and water vapour (GHGs) on the atmosphere. Warming has increased rapidly in the past 200 years as a result of human activity and GHG emissions. This is known as the enhanced greenhouse effect.

8 Clouds reflect sunlight back to space leading to cooling, but clouds can also restrict heat radiated back, leading to warming. High clouds retain heat longer, low-level clouds reflect more sunlight. In a warming world we may get more high cloud.

9 Rapid warming in the Arctic is due to regional albedo. As ice (which is highly reflective) melts, it exposes land and ocean surfaces that absorb more of the sun's radiation. The result is a rapid rise in average temperatures.

10 Global warming in Africa will mean longer and more intense droughts and more flooding. Large parts of Africa are semi-arid and small reductions in rainfall in these areas could make farming unsustainable. The poverty in Africa also means that people do not have the resources to cope with climate change.

11 Trees reduce carbon dioxide emissions to the atmosphere by storing carbon above and below ground. They also absorb carbon dioxide from the atmosphere during photosynthesis.

12 Ocean fertilisation stimulates the growth of phytoplankton which absorb carbon dioxide.

13 Many countries in some way contribute to climate change, the effects spread globally and so it is a global issue. Also the atmosphere is part of the 'global commons'.

14 (a) The Kyoto Protocol is the first legally binding international agreement on climate change.

(b) Carbon trading is a market-based system aimed at reducing GHG emissions.

(c) Carbon credits are a certificate or permit which represents the right to emit 1 tonne of carbon dioxide; credits can be traded for money.

Option B Disease dilemmas

15 Infectious disease is spread by pathogens, e.g. bacteria and viruses. Contagious disease is a class of infectious disease easily spread by direct or indirect contact with people.

16 The disease spreads through a structured sequence of locations from large, well-connected centres to small isolated centres.

17 Primary; diffusion; condensing; saturation.

18 Climate affects where vectors of disease live, e.g. mosquitoes in humid tropics; low levels of sunlight can lead to diseases derived from deficiency of vitamin D; cold, damp climates can lead to a prevalence of respiratory diseases.

19 Altitude: with increasing altitude, temperatures fall, e.g. there are fewer mosquitoes with increasing altitude; on flat floodplains, standing water can become polluted and spread disease in LIDCs.

20 Epidemics of influenza and respiratory illness peak in the winter months in the Northern Hemisphere. Vector-borne diseases transmitted by mosquitoes, flies and ticks peak during the rainy season in the tropics and sub-tropics. Fly populations are highest in South Asia in the pre-monsoon (March–April) and end of monsoon (September–October) seasons. The tsetse fly, which transmits sleeping sickness in west and central Africa, can live longer in the wet season. Freshwater snails, which transmit bilharzia to humans, follow a seasonal lifecycle pattern linked to precipitation and temperatures of 10–30°C.

21 A model suggesting that over time, as a country develops, there will be a transition from infectious disease to chronic and degenerative diseases as the main cause of death.

22 Classical/western, e.g. western Europe; accelerated, e.g. parts of Latin America; contemporary/delayed, e.g. sub-Saharan Africa.

23 A disease barrier can interrupt the spread of disease; it may be physical, e.g. mountains, or human, e.g. border controls.

24 Physical barriers isolate communities and reduce the spread of disease. Physical barriers restrict population movements and therefore limit the spread of a disease.

25 An epidemic (a disease outbreak spreading quickly through a population) that spreads worldwide, e.g. Spanish flu.

26 Over-harvesting, which reduces plant populations and genetic diversity. Habitat destruction, e.g. deforestation in the tropics.

27 It describes a disease that exists permanently in a geographical area or human group.

28 Top-down: large-scale operations relying on the input of international organisations such as the World Health Organization and governments. Bottom-up: local level projects, also called 'grass-roots strategies', which involve and empower local communities.

29 Because they can engage with and relate to women more effectively and women are primary carers of children and usually have a primary role in hygiene in the home. They can also understand a particular role undertaken by women that can spread disease, such as sourcing drinking water.

Option C Exploring oceans

30 Salinity affects water density and water density in turn affects movements such as ocean currents.

31 Surface currents and deep water currents.

32 This is a continuous, slow-moving flow of ocean currents around the globe. It is explained in the following sequence:

1. Most oceans hold warm water at the surface and cold water in their depths.
2. Surface water tends to move from low to high latitudes.
3. In polar regions ocean surface currents are cooler, more saline and denser.
4. Water sinks and moves horizontally.
5. Deep currents flow back to the equator where water rises.

33 The layer of the ocean where photosynthesis is most active — generally the light zones nearer the surface as light decreases with depth.

34 Nutrient levels are relatively low at the surface and are quickly used up. Cold, dense, deep water holds nutrients nearer the ocean floor. Some exceptions include the Southern Ocean and Antarctica where a strong upward movement of water moves nutrients to the surface.

35 The Earth's shared natural resources, e.g. oceans and atmosphere.

36 This is where people exploit a resource and take all that they can, believing that the cost will be shared among a much larger number of people. This can lead to depletion and degradation of the resource.

37 A system of oceanic circulation which transports and distributes pollution over vast distances and can lead to areas of accumulation.

38 Ocean acidification and ocean warming.

39 Eustatic: absolute changes in sea level. Isostatic: changes in the absolute level of the land.

40 Causes of sea level change include increase in surface water temperatures: sea water decreases in density, volume increases and sea level rises. This is called **thermal expansion of water**. Melting of glaciers and small ice caps due to increase in air temperatures. Melting of Greenland and Antarctic ice sheets. This is more complex as the central parts are thickening slightly but the edges are thinning rapidly.

41 More trading of goods across the oceans; mass movements of people; more movement of resources and more exploitation of resources as countries develop; more movement of people through tourism; more interconnectedness, e.g. fibre-optic cable laying.

42 Disputes arise when there are known reserves of energy and mineral resources in the area. There may be disputes to exert political influence by seeking to extend marine borders.

43 Poverty; organised crime; dysfunctional governments; loss of traditional fishing areas.

Option D Future of food

44 A commonly used definition of food security comes from the United Nation's Food and Agriculture Organization (FAO):

Food security exists when all people, at all times, have physical and economic access to sufficient, safe and nutritious food that meets their dietary needs and food preferences for an active and healthy life.

45 Availability; access; utilisation; stability.

46 It contains minerals and nutrients essential for plant growth; it supplies water and nutrients to the plants' root systems.

47 In warm climates rainfall evaporates quickly, so reducing its effectiveness.

48 New trade routes and improved access to global food sources; changing global tastes because of a wider sharing of ideas and food tastes.

49 Globalisation has influenced dietary patterns. With increased affluence, consumption of food has shifted away from cereals and towards meat, dairy, more processed foods and fast foods. Fast-food outlets promoted by TNCs have increased. These changes, together with a more sedentary lifestyle, lead to obesity.

50 Slope angle affects rates of soil erosion, use of machinery and soil formation (depth and drainage). On steep slopes soils are often thin, poorly developed and excessively drained. Soils at the base of a slope can become waterlogged. On gentler slopes, there is less erosion and leaching.

51 The reduction of agricultural capacity due to over-exploitation of resources and natural processes such as drought. In extreme cases it results in desert-like conditions.

52 Population growth increases food demand; a combination of overgrazing, expansion of cropped areas and more intensive agriculture causes land degradation; and taking marginal areas into production impacts on production as soils are already fragile on marginal land.

53 Reduced habitats; increased competition and population decline of some species; loss of biodiversity; food webs become fragile.

54 Poverty → increased pressure on the land to produce food → overgrazing and more intensive farming → reduced vegetation cover → increased soil erosion → reduced land productivity → poverty.

55 Warming of the atmosphere increases the number of times that temperatures reach extreme levels, more water vapour evaporates from the oceans → an increase in water vapour in the atmosphere → more intense rainfall → the retention of heat energy from the sun → further global warming.

56 The volume of water needed to make a product or grow a crop, measured at the place where the product is made or the crop is grown.

57 It leads to a process of eutrophication: nutrient enrichment in water leads to a fall in oxygen levels and the death of species which are oxygen dependent, affecting food chains and food webs in freshwater ecosystems.

58 ACs and EDCs are both at risk of increasing levels of obesity as it is connected to rising affluence and a more sedentary lifestyle. Dietary changes from cereals and vegetables to dairy, meat and processed food lead to excess sugar and calories in diets.

59 Economic recession; food supply shocks; civil unrest over food shortages; reduced food supply as a result of global warming.

60 Donor-driven food aid centres on the use of food aid as a means of 'dumping' surplus food from ACs. Food aid dependency could be a long-term outcome of food aid for the recipient countries. Large quantities of food aid can swamp local markets and drive down prices, reducing the income of indigenous farmers.

61 Costs: poor working conditions; exploitation of resources needed for indigenous food production; use of the most productive farmland; marginalisation of small-scale farmers; negative environmental impacts. Benefits: job creation; increased income which can generate the multiplier effect; reduction of poverty; bring in new technology; improve farming skills.

Option E Hazardous Earth

62 The Earth consists of the core, mantle (asthenosphere and lithosphere) and crust. The core is the centre of the Earth, with an inner core 1,250 km thick and an outer core 2,200 km thick. The mantle forms 80% of the volume of the Earth and is semi-solid rock. The crust is the outer shell and consists of oceanic and continental crust.

63 Oceanic crust comprises basalt, is relatively thin (5–10 km) and is less than 200 million years old. Continental crust is made of granite, is up to 70 km thick and may be several billion years old. Continental crust is less dense than oceanic crust.

64 Polarisation of iron particles in the oceanic crust. Periodic reversal of the poles causes iron particles in molten rock to be aligned alternately towards the north and south poles. These patterns are mirrored on either side of the mid-Atlantic ridge, proving that new crust formed here and spread away from the ridge.

65 As lava erupts it cools and the magnetic orientation of iron particles within the lava is 'locked' into the rock. The direction of the Earth's magnetic field changes every 400,00–500,000 years (palaeomagnetism).

66 Explosive: violent because of build-up of pressure, viscous magma (e.g. andesite) which prevents escape of gases; effusive: a gentle, free-flowing basic eruption of lava, e.g. basalt. Check your points against Table 4.30 on p. 204.

67 Lava flow (hot, liquid rock); pyroclastic flow (fast-moving flow of hot ash, boulders and gases); lahar (mudflows); ash fall (tiny rock fragments and pumice emitted into the atmosphere).

68 A process by which sediments and soils lose their mechanical strength from a sudden loss of cohesion.

69 A wave trough formed in front of a tsunami where sea level is reduced.

70 Shallow-focus earthquakes: surface down to approximately 70 km. Often occur in brittle rocks. Generally, release low levels of energy but high-energy shallow quakes can cause severe impacts. Deep-focus earthquakes: 70–700 km. Increasing depth leads to high pressure and temperature. Less frequent but very powerful.

71 Primary hazards: ground shaking and liquefaction. Secondary hazards: tsunamis, landslides, debris flow, fires.

72 Construction of earthquake-proof buildings and infrastructure (building codes, deep foundations, reinforced structures, fire-resistant structures). These measures are costly and can be beyond the resources of LIDCs. Disaster planning to provide emergency relief, a plan for long-term recovery. These measures can also be addressed through international aid.

73 The number of natural disasters has increased over time but the increase in earthquakes and volcanic eruptions is less pronounced. There has been an increase over the past 50 years; on average there are now 30 volcanic eruptions and earthquakes a year. Data show that earthquakes have a significantly greater impact in terms of number of deaths than volcanic eruptions. Generally, the greater the magnitude of an eruption or earthquake, the less frequently it occurs.

74 Two of these strategies:

- Identification of a specific and reliable 'precursor' (a characteristic pattern of seismic activity or a physical, chemical or biological change) in order to confidently predict an earthquake event within a specific time and place.
- Future research on improving the forecasting of earthquakes.
- Alleviate poverty to reduce the impact of earthquakes in the future.
- Continued advances in building design to withstand tectonic events, as falling buildings account for the highest proportion of deaths and injury.

75 Natural hazards exist at the interface between the physical and human environment. A hazard becomes a disaster when there is loss of life and/or destruction of the built environment, and/or disruption to human activities.

76 Its magnitude and duration. It is also determined by a range of environmental, social and economic factors such as mitigation, experience, perception, preparation, physical context, technology for warnings and response, wealth and vulnerability.

77 Socio-economic status, education, employment, culture, past experience, values.

78 Response, resilience, prediction, protection and prevention.